MIKE

DAVID G. BAKER

MI

THE TIGE

KE

RS OF LSU

Louisiana State University Press • Baton Rouge

Published with the assistance of the V. Ray Cardozier Fund

Published by Louisiana State University Press
lsupress.org

Copyright © 2023 by Louisiana State University Press
All rights reserved. Except in the case of brief quotations used in articles or reviews, no part of this publication may be reproduced or transmitted in any format or by any means without written permission of Louisiana State University Press.

Manufactured in Canada
First printing

DESIGNER: Michelle A. Neustrom
TYPEFACES: Termina, display; Whitman, text
PRINTER & BINDER: Friesens Corporation

COVER PHOTOGRAPH: Mike VII on rock, courtesy of Bobby Gilboy
DEDICATION PAGE: Mike VII in pool, courtesy of Katie Miley
OPPOSITE ACKNOWLEDGMENTS: Mike VII with plants, courtesy of Katie Miley

LIBRARY OF CONGRESS CATALOGING-IN-PUBLICATION DATA
Names: Baker, David G., 1956– author.
Title: Mike : the Tigers of LSU / David G. Baker.
Description: Baton Rouge : Louisiana State University Press, 2023.
Identifiers: LCCN 2023007848 | ISBN 978-0-8071-8030-3 (cloth)
Subjects: LCSH: Louisiana State University (Baton Rouge, La.)—Mascots—History. | Mike (Tiger)
Classification: LCC LD3116 .B35 2023 | DDC 378.763/18—dc23/eng/20230411
LC record available at https://lccn.loc.gov/2023007848

To my wife, Brenda, who for twenty-six years patiently encouraged and supported my care of Mike the Tiger, which included adjusting our family time to accommodate Mike's needs.

CONTENTS

ACKNOWLEDGMENTS

I would like to acknowledge the amazing group of veterinary students who provided Mike's day-to-day care under my supervision and mentorship. Their dedicated, conscientious, and loving care of our tigers foretold equally successful careers as practicing veterinarians. Perhaps the most enjoyable and rewarding aspect of working with Mike the Tiger was mentoring his veterinary student caretakers.

I would also like to acknowledge my friends and colleagues Dr. Rhett Stout, who quietly and selflessly served as Mike's back-up veterinarian for twenty-six years; and Dr. Gordon Pirie, former veterinarian for the Baton Rouge Zoo, who, on many occasions, graciously assisted me with Mike's care and to whom I could always turn with questions. My care of Mike the Tiger was greatly enhanced by the contributions of these two professionals.

MIKE

MEMORIES OF MIKE

"Let me show you my cat." I thought, "Sure, it's not like I haven't seen a cat before." I was touring LSU for the first time. My host was Dr. W. Sheldon Bivin, attending veterinarian for LSU–Baton Rouge and director of the Division of Laboratory Animal Medicine (DLAM) in the School of Veterinary Medicine (SVM). I was interviewing for the position of chief clinical veterinarian in the DLAM. The division provides husbandry and veterinary medical care for animals used in teaching and research on the LSU campus. Because I was trained as a laboratory animal veterinarian, and not as an exotic animal veterinarian, I was quite surprised to discover that the "cat" my host referred to was in fact a nearly four-hundred-pound tiger on a university campus. In addition to his other duties, Dr. Bivin served as the veterinarian for Mike V, LSU's live tiger mascot. Upon seeing Mike, my first thought was, "Only in the South!" Little did I know that eleven months after joining the SVM faculty, I would take over responsibility for Mike V when Dr. Bivin retired from LSU.

With Dr. Bivin's departure, I became attending veterinarian for LSU, DLAM director, and director of the SVM's Raptor & Wildlife Rehabilitation Unit, in addition to other duties. As soon as it became apparent that Dr. Bivin was leaving LSU and that I might take on his responsibilities, I began visiting Mike V so that he could get to know me and I could become familiar with his routine. On July 1, 1996, I was given the keys to the tiger enclosure and formally became Mike the Tiger's third veterinarian. It was a steep learning curve. There hadn't been time for much of a transfer of information, so I had to figure Mike V out on my own. Within a short time, I was extremely

opposite Checking Mike VII's teeth.

Feeding Mike V a treat while I examine him. (Harry M. Cowgill)

fortunate to come to know Dr. Gordon Pirie, then of BREC's Baton Rouge Zoo. Over the years, Dr. Pirie became a close friend and valuable resource to whom I would turn with questions about tiger care.

A Close Call with Mike V

Mike V was six years old when I assumed responsibility for his care. He and I came to know each other quickly. I drove an old pickup truck at the time, and Mike learned to recognize the sound of my truck door closing. When he heard it, he would look up from his nap and watch for me. I had never worked with exotic cats, nor many other cats for that matter. Regardless, over time I was struck by how similar Mike V was to domestic house cats in the way he behaved and even in his physiologic response to drugs and other medications. That is why I was surprised, early in my tenure as his veterinarian, when I nearly killed him with a cocktail of drugs I had successfully administered to several house cats without incident.

One of the claws on Mike's foot had become overgrown and was piercing his paw. It was very painful, and he was unwilling to allow me to examine it. In order to do so, I sedated him with a common drug combination. Unfortunately, he began to seizure, a reaction only rarely observed with that drug combination when used in house cats. I quickly called my associate, Dr. Rhett Stout, who brought over a dose of valium. Valium is the antidote for seizures. After administering the valium, I watched Mike V for several minutes after the seizures stopped, then left to do my daily workout. When I returned, I stood in Mike's night house, observing him. His chest was not moving; he wasn't breathing. I thought, "Wow, I killed him." But, to be sure, I smacked him on the chest. He took a deep breath and, thankfully, continued breathing. My next thought was, "That was close!" Needless to say, I didn't use that sedative combination again. Only after a few years did I come to appreciate how fortunate it was that I hadn't killed Mike the Tiger!

If I had to describe the personality of Mike V, I would say he was calm and stoic—at least usually. A few times I had to force him into his travel trailer for a ride to the SVM for medical care. As an extremely intelligent animal, he always knew when it wasn't a football game day and usually refused to get into the trailer, uncertain of the reason or destination. On some occasions, as I pushed on a pivoting wall of chain-link fencing, trying to force him into his trailer, he became very agitated with me, roaring at me and attacking the fencing while I was just inches away on the other side. The roar of a tiger in a small, enclosed space is awe-inspiring, and terrifying!

Mike V lying in grass in his new enclosure. (Harry M. Cowgill)

Roscoe at five weeks old.
(Donna Cotner)

Except for those rare situations, Mike V was very interactive. He usually came right over to me when I visited and allowed me to scratch him and to palpate various lumps and bumps on his body. It was in doing so that I once discovered a tumor on his abdominal wall, which I removed surgically with the help of Dr. Pirie. Mike V and I had a good relationship, even if not as close as the relationships I would later have with Mikes VI and VII. Back then we were still taking Mike V to the ride-around prior to football games. If Mike became a bit nervous on the field, he would lift his head and look around until he saw me. After reassuring himself that with me there all must be well, he would lay his head back down and relax.

The death of Mike V in May 2007 was hard on all those who loved him. However, I was confident that we had provided Mike V with the best available care. While I have certainly enjoyed the relationships I've had with the tigers, my primary feelings toward them are as a caregiver. I have always taken very seriously the responsibility for the tiger's care. He can't provide that for himself in captivity, so we must serve as his greatest advocates, providers, and protectors. By the time Mike V died, I had become very comfortable with my role as his veterinarian and had grown tremendously in my knowledge of tiger behavior and care.

Enter Roscoe

After the death of Mike V, I confirmed with Chancellor Sean O'Keefe that I should find a new tiger mascot. Having cared for Mike V for nearly eleven years, I had a pretty good idea of what to look for. I wanted a tiger that was confident, interactive, inquisitive, in good health, well-proportioned and pleasing in appearance, with a healthy genetic and medical background, and without any legal baggage that might embarrass LSU or embroil the university in litigation. I also wanted a young tiger so that, hopefully, he would be mascot for many years. However, he couldn't be too young because neither I nor my staff had time to hand-raise a cub. After many calls to colleagues in the field, I learned of a tiger cub named Roscoe at the Great Cats of Indiana facility. Dr. Pirie accompanied me to examine several young male tigers from two litters. We both agreed that Roscoe had all of the desired characteristics and was perfect for LSU.

Bringing the next Mike to LSU was very important to the LSU community. PETA and other extremist animal rights groups had done all in their power to discourage LSU officials from continuing the live mascot tradition. Many within the LSU community were concerned that there wouldn't be another live tiger on campus. However, as Mike VI was transported into Tiger Stadium for his first football game a few days after his arrival, people lined the route up to twenty deep. Some people were cheering, others were crying, while others were standing silent with their hands over their mouths. In that moment he wasn't just a tiger; he wasn't even just Mike the Tiger. In that moment he was their tiger. Dr. Pirie was with me for that first game. When we entered the stadium, we turned to each other and nearly simultaneously said that we had never experienced anything like it before. The emotion was almost palpable. It was a reminder to me of what Mike the Tiger means to the LSU community and the responsibility I bore as his primary caretaker.

Of the three LSU tigers I cared for, Mike VI was my favorite. He was

Mike VI's first football game, October 6, 2007, against the University of Florida. *Upper left:* loading into his trailer (Wendy Day); *upper right:* relaxing before the game (Wendy Day); *left:* receiving water from caretaker Wendy Day. Leaning on the truck in the background in gray and white t-shirts are Mr. and Mrs. Rob Craig, who donated Roscoe to LSU (Ginger Guttner).

Playing with the boomer ball. (Christine Russell)

very affectionate and interactive. Even as he aged, he never completely lost his youthful playfulness. It also seemed to me that he was the most responsive of the tigers. Once, while I was playing with him, he managed to bite my arm, drawing blood and causing me to wince. He immediately adopted a sad expression and began licking my arm. He seemed to be remorseful for hurting me. Needless to say, a tiger like that is a pleasure to care for.

Because of his wonderful personality, it was particularly difficult on his entire care team when Mike VI developed terminal cancer at the relatively young age of ten years. His veterinary student caretakers were especially affected. The student caretakers serve for two years, during their third and fourth years of veterinary school. Mike's caretakers were about to graduate when one of them observed very mild asymmetry in his face. This would later become more obvious and would be diagnosed as cancer in the bones of his face. Such a finding just days before relinquishing responsibility for Mike VI was very hard on the outgoing students. And it was no easier on the new team of student caretakers. Imagine taking over the care of a beloved animal with a terminal illness. The students didn't know how to deal with Mike's condition and asked, "How do we function as a veterinarian under these conditions?" I encouraged them to watch me and learn from my example. While Mike's cancer and ultimate death were very hard on all of us, I considered it a great honor to serve as a role model for my students.

The familiarity Mike VI and I shared was evident when his cancer returned after treatment and his clinical signs worsened. He would not enter his night house for several days. As his veterinarian, I was concerned that we would not be able to provide him with the medical care he needed if he remained out in his habitat yard. Cats are very good at masking their illnesses. So if a cat appears ill, it is often very ill. The worst thing that could happen for Mike VI and for LSU would be for Mike to languish and die in his yard. He needed to come inside so that he could be evaluated and, if necessary, humanely euthanized. So when Mike's student caretakers phoned and told me that he wouldn't enter his night house, I came down to campus, carefully entered his yard, and called to him. He slowly arose and came inside the night house. A few days later, I humanely euthanized Mike VI. Later in this book I tell of Mike's final goodbye to Caitlin Bell, one of his veterinary student caretakers. It was a special moment for all of us. I had promised the LSU community that Mike VI would not suffer but would have a "good death." His was about as good a death as one could hope for.

Mike VI with post-radiation eye patch. (Christian Folk)

It was during the reign of Mike VI that we discontinued the pregame ride-arounds in Tiger Stadium. The Athletic Department had begun launching fireworks next to the tiger trailer, causing Mike VI to refuse to load into his trailer. Our decision to stop the ride-arounds caused quite an uproar. I have always said that there is nothing special about me as a person or a veterinarian, and that any competent veterinarian could provide adequate care for Mike. However, one characteristic essential in Mike the Tiger's veterinarian is the ability to say no and mean it. I became the target of many hurtful and false accusations for refusing to force Mike into his trailer. Fortunately for both Mike VI and LSU, I just didn't care. Mike's well-being was all that mattered to me.

A New Tiger on the Scene

After the death of Mike VI, we postponed our search for a new tiger to allow for some needed renovations of the tiger habitat. Then, in August 2017, Drs. Stout and Pirie assisted me in bringing 11-month-old "Harvey" to LSU from a facility in Florida. While many in the LSU community had said they could never love another tiger like they loved Mike VI, the new tiger quickly won their affection. My plan, as I explained to the LSU administration, was to bring a young tiger to LSU to help the university community get past the

loss of Mike VI. It seemed to work. Because there were no longer plans to take Mike VII to the pregame activities, caring for the tiger became less about public interactions, as it had been for some of the previous student caretakers, and more about the tiger and his care. This was good for Mike. While all of the previous student caretakers had been extremely responsible and attentive to Mike's needs, now it was their only concern. Public recognition would no longer be a distraction.

My relationship with Mike VII was very different from my relationship with Mike VI. Mike VII was, and has remained, much more aggressive. He too once bit my arm and drew blood. Rather than showing signs of remorse, he appeared quite pleased with himself and as though he wanted to do it again. Nevertheless, he is usually affectionate and interactive, if less consistently than the two tigers I cared for before him. He also appears to have less tolerance for annoyance by his caretakers. While previous tigers would usually allow me to vaccinate them as they leaned against the caging in the night house, Mike VII wants no part of it. As soon as he feels the needle, he bolts! He will likely need to be restrained for his vaccinations for his entire life.

Elsewhere in this book I tell the story of how Mike VII seems to need physical contact with his caretakers. That was a surprise to me. I knew that tigers liked human contact with those they know, but I hadn't anticipated the importance of those interactions. As I stated previously, tigers are in many ways like domestic house cats. Life is on their terms. One minute they want your attention and affection, while the next minute they want nothing to do with you. Over the years, many people have asked whether I am a "cat person." While I am not, caring for Mikes V, VI, and VII has enabled me to understand cats better and, through that understanding, appreciate them more. But I still don't plan to get a house cat!

Serving as veterinarian for LSU's Mike the Tiger has been the opportunity of a lifetime. I learned much and met many wonderful people along the way. I am certain that, like my veterinary students, I have changed much through the experience and, hopefully, am now a better professional than I would have otherwise been. I am thankful for the support of the Athletic Department and the LSU administration as I have striven to provide Mike with the best care possible. Geaux Tigers!

above Visiting with Mike VII. (Burton Landry)

opposite Mike VII gives the camera a serious look. (Katie Miley)

CHAPTER I

(2017–PRESENT) MIKE VII

In 2017, after the difficult death of Mike VI from cancer, I consulted with the LSU administration. Chancellor F. King Alexander confirmed that the university would acquire another tiger. The timing of this tiger search would be a bit different, however. With the tiger habitat unoccupied, we had the opportunity to perform some much-needed renovations and improvements. After a complete assessment, we set about upgrading the enclosure. We had the floor inside the night house refinished and resting platforms installed, to enable Mike to get up off the floor. This preference is shared by many cat species. Outside, workers removed and replaced the entire pool and stream system to allow for repair of some pesky leaks. We had the metal pole supporting the stainless-steel mesh canopy modified with concrete to give it the appearance of a tree. In fact, the façade now looks so realistic that guests regularly ask if Mike ever climbs his "tree." Perhaps most important to Mike, we installed a resting platform, in the form of a flat rock, in the yard. The platform contains coils that are cooled in the summer and warmed in the winter. Mike can often be found lounging on his rock. These upgrades to the enclosure greatly improved Mike's comfort and care.

Finding a New Tiger

As renovation of the tiger habitat got under way, I began the search for Mike VII. I sent out queries through government officials and others familiar with the US captive tiger population. These government agencies and friends in the sanctuary community began watching for a suitable tiger. During

opposite Mike VII relaxes on his rock. (Bobby Gilboy)

the search, I learned of Animal Adventures, a Florida sanctuary that was in some regulatory hot water. The federal government had found the facility in noncompliance on certain animal care regulations, with the result that the sanctuary had lost its exhibitor's license. This is a common occurrence since many facilities take in, with the best of intentions, more animals than they can effectively manage.

Initially, I was not interested in dealing with a facility that was experiencing regulatory problems with the US government. But after some thought, I realized that such a facility might have a tiger that truly needed a new home—that is, a rescue animal that would be a good fit for LSU. As part of a plan forward with government officials, the facility changed ownership and was renamed the Wild at Heart Wildlife Center. With the encouragement of state and federal wildlife agents, I contacted the new owners, Jeremy and Jaime Hargett, and learned that they were making a serious effort to bring the facility back into compliance. Their efforts included downsizing their collection of tigers. They invited me to visit and examine a young tiger named Rocky. So, in June of 2017, courtesy of the Tiger Athletic Foundation, I flew to Florida with Dr. Gordon Pirie, who had accompanied me to Indiana on the search for Mike VI.

Dr. Pirie and I examined Rocky. While the 15-month-old tiger would have been acceptable, he had two physical features that caused me a bit of concern. First, his facial conformation gave the appearance that his eyes were crossed. Second, his testicles were small. Neither of these features mattered to me, but if possible, I didn't want to give fans from other schools something to criticize about the next Mike the Tiger. Almost as an aside, we were also introduced to Harvey, a 9-month-old Sumatran mixed-breed tiger cub, along with his sister, Haley. It was quickly apparent that Harvey had all the characteristics needed to become LSU's next mascot. He was a healthy, good-looking tiger with an engaging, confident personality. In addition, he desperately needed a permanent home. The Hargetts graciously offered to donate Harvey to LSU.

With a tiger selected, I made plans over the course of the next several weeks to bring him to LSU. We needed to be sure that Harvey was healthy for travel, and I needed to obtain various permits to transport him across state lines. So I arranged with Dr. Jonathan Shetler of Sunrise Animal Hospital in Sunrise, Florida, to perform a complete physical examination and to administer the appropriate vaccinations.

Harvey explores his night house after arriving at LSU. (*left:* Caitlin Bell; *right:* Ginger Guttner)

On August 15, 2017, we again set off for Florida. In addition to Dr. Pirie, I was accompanied by Dr. Rhett Stout, Mike's longtime back-up veterinarian. Richard Zuschlag, owner of Acadian Ambulance Service, offered the use of one of the company's aircraft, as he did when we picked up Roscoe (Mike VI) in Indiana in 2007. Michael Sonnier, Acadian's chief of flight operations, worked out the flight details, and pilot Spencer Anderson flew the team to Florida, as he had done when LSU brought Roscoe from Indiana. It was comforting for the tiger team to work with partners they knew and trusted. The trip could not have gone better, due in large part to competent help from many people, as well as Harvey's very tractable nature.

After arriving at LSU, Harvey underwent a quarantine and acclimation period. This allowed him to adjust to his new diet and environment, receive preventive and therapeutic medical treatments, and have the opportunity to bond with his new caretakers. On August 21, 2017, he was released into his yard and, at that moment, became Mike VII.

As a veterinarian, I found it fascinating to observe the growth and development of a tiger. I was particularly interested to watch Mike's adult teeth appear and replace his "baby" teeth, perhaps because the huge canine teeth of an adult tiger are among their most impressive features. It was also fascinating to watch his musculoskeletal development. Mike grew from a

left Mike VII in September 2017. Notice the size of his paws in relation to the rest of his body.

right Enjoying his yard.

opposite A young Mike VII explores his new home. (Eddy Perez, LSU)

somewhat gangly young tiger into a well-muscled adult. I had always known that, as apex predators, tigers are physically imposing animals, but having never been in close proximity to an adult tiger before my time at LSU, I really couldn't appreciate just how powerful they are. It is no wonder that they are such effective hunters.

Along with growth in his musculoskeletal system came a deepening of Mike's coloration. As a young tiger, Mike was much lighter on the front end of his body than on the back end. Over time, the orange fur and black stripes intensified, resulting in a really striking appearance.

Finally, it was also interesting to observe the maturation of Mike's personality. In many ways it was much like observing a young person mature into an adult, leaving behind the interests of youth and taking a more serious view of life.

The COVID-19 Pandemic

In early March 2020, as the COVID-19 pandemic was spreading across the United States, I instructed Mike's student caretakers to stop touching him and to wear gloves and masks when working in the night house. It seemed prudent to take this precautionary step even though at that time no cats of any species, including tigers, had been found infected with the SARS-CoV-2 virus, the cause of COVID-19. I also asked the Athletic Department to erect a barrier in front of the tiger enclosure to keep people at least ten feet away from Mike VII. The week after I made this request, a tiger at the Bronx Zoo in New York died of COVID-19, and shortly thereafter house cats were also found to be susceptible. These findings confirmed the importance of taking steps to protect Mike from the virus.

Unexpectedly, over the course of the following several months, Mike VII developed signs compatible with a condition known as feline idiopathic cystitis (FIC). While FIC is common in domestic cats, I was unaware of any reports in tigers or other large exotic cats. Regardless, the clinical signs fit. Of greatest concern was the observation that Mike's urine would occasionally turn red. This coloration indicated the abnormal presence of hemoglobin, a blood protein, and occasionally the presence of small numbers of red blood cells. I made attempts to resolve the condition with medical treatment, but nothing worked. In addition to the change in Mike's urine, his behavior changed slightly as well. He seemed to become a bit more "needy." That is, he seemed to want to be near us even more than before the pandemic.

After several months, I concluded that the lack of human contact was proving stressful to Mike VII and may have caused the FIC as well as the change in his behavior. By this time all of Mike's caretakers were vaccinated, so we decided to resume physical contact with Mike. Almost immediately, his attitude returned to pre-pandemic normal, and the FIC resolved. Apparently, physical contact with his caretakers was more important to Mike than I or anyone else had appreciated.

A few months later, I obtained a SARS-CoV-2 vaccine developed for use in animals and administered two doses to Mike VII. Mike was not happy about getting stuck with a needle, but it was for his good. It has been demonstrated that large exotic cats are highly susceptible to SARS-CoV-2 and have an unacceptably high mortality rate following infection. While there have been a few break-through cases of infection in fully vaccinated tigers, clinical signs tend to be less severe following vaccination. The COVID-19 vaccine will be added to Mike's routine annual vaccination schedule.

After twenty-seven years I retired from LSU and took a position in private industry. It was a very positive transition in my career. However, as might be expected, it was difficult to relinquish care of Mike the Tiger and mentoring of his veterinary student caretakers. The transition was made a

Something catches the attention of Mike VII. (Henry Hebert)

bit easier by turning over Mike's care to Dr. Stout, my long-time colleague and friend, who is not only a highly competent veterinarian, but also very familiar with Mike and the entire mascot program. Dr. Stout had assisted me with all aspects of the program for the past twenty-six years. With complete confidence in Dr. Stout, I was able to leave Mike VII knowing that he is in good hands and will continue to be well cared for.

MIKE VII'S STUDENT CARETAKERS

- 2017–18: Caitlin Bell and Christian Folk
- 2018–20: Nick Fleissner and Robert ("Tré") Gildersleeve
- 2020–22: Katie Miley and Cora Schneider
- 2022–present: Erin Guillory and Amelia Talbot

REMEMBRANCES FROM

MIKE VII'S CARETAKERS

Christian Folk (2017–18): Christian recalled how young Mike VII created such a mess in his night house. Each morning, Christian and his partner found that Mike had scattered his cedar shavings throughout the night house. Perhaps after sleeping on his shavings, he decided it was time to play with them! Christian also remembered how Mike VII consistently dipped his oxtail treats in his water bowl prior to eating them. Why he did this is anyone's guess. He may have done it to warm them up a bit, since they were given to him directly from the refrigerator. Or perhaps he just liked them very clean or very wet. Lastly, Christian recalled how Mike VII seemed to enjoy looking in the window of his night house while the students were inside cleaning, as if he was inspecting their work! *Dr. Folk is a small-animal surgery resident at Cornell University in Ithaca, New York. He intends to practice as a board-certified veterinary surgeon.*

Nick Fleissner (2018–20): "I remember the time we hung a boomer ball in Mike's habitat. It seemed quite complex for an enrichment toy with the ball, chains, a heavy-duty spring, PVC pipe, fire hose, clamps, etc. We hung it from the top of his enclosure, and he seemed to love it. About two weeks later I came to his enclosure one night to find him running around, obsessed with something in the stream. It took a while, but I was finally able to get him to come inside to eat and sleep for the night. I did not look in his yard that night because it was dark, there were people at the fence, and I figured I would be out in the yard the next morning as usual before letting him out anyway. The next morning, I found the boomer ball at the bottom of the stream. I had to straddle the stream and pull the already heavy boomer ball out, now full of water, knowing I was being recorded the entire time on the livestream camera. I didn't want to fall into the stream!

"It took a couple of months, but we were able to repair the toy. We rehung it one evening. This itself was quite a task as one person had to be on the top of the enclosure to lift the chain and ball up and secure it to the canopy netting while someone else supported it from the ground. We let Mike out the next morning, and he seemed incredibly excited. A few hours later when I was in class, I decided to check the livestream camera to see if he was enjoying his toy. Immediately I noticed that the hanging boomer ball was not there. The livestream camera allows you to see the last twelve hours, so I was able to rewind the tape and see what happened. It took only a couple of hours for Mike to demolish all the work we did the night before and in the months leading up to it. Even at the young age of 2.5 years, Mike had such incredible power that dragging the boomer ball into the bushes created enough tension to snap the hanging support. It was quite amazing to see on camera, making it hard to be mad at him.

Nick Fleissner hangs the boomer ball for Mike VII.

"At lunch I headed over to the enclosure to let Mike in his night house so I could clean up the carnage of the toy. We did not attempt to fix it again after that. We knew he was growing quickly and could tell he was powerful just by his appearance, but I'll never forget all the work and planning we did to place that toy in his enclosure twice, and ensure it was durable, just for him to rip it down within a few hours." *Dr. Fleissner is receiving specialty training in veterinary anesthesia and pain management at the University of Wisconsin School of Veterinary Medicine in Madison, Wisconsin.*

Katie Miley (2020–22): "Caring for Mike VII was different from any other animal care experiences I have had. His emotional intelligence is far superior to that of any other animal I have worked with. He is attentive to his caretakers. He learned the sound of my footsteps and specifics about me that alerted him to my presence before I was even in view. He recognized my body language and attitudes and notably changed his behavior in relation to mine.

"While reading about tigers prior to being selected for the role of caretaker, I happened upon an article that correlates caretaker attitude and outlook with tigers' well-being and overall happiness. After two years of caring for Mike VII, I can confirm that, at least in his case, it was 100 percent true. I kept that article in mind throughout my tenure as caretaker and made a conscious effort to enter his presence with gratitude for this role. There were only a handful of days on which he did not enter the night house for me in the evening, and I found that most of those nights I had a bad attitude, except for a few large sporting events in which the crowd interested him more than dinner.

"It has been one of the greatest joys in my life to know a tiger so intimately as to recognize his mood with a glance and in turn have him understand mine. He often ran to greet me like an old friend, and I strove to greet him with the same enthusiasm. The relationship I developed with Mike VII is one I will remember for the rest of my life. I am a Baton Rouge native and so return often, and always visit him." *Dr. Miley is an equine practitioner.*

Dr. Katie Miley as a newly graduated veterinarian says goodbye to Mike VII. (Erin Whitney)

Cora Schneider applies preventive medication to Mike VII. (Katie Miley)

Cora Schneider (2020–22): "I began training to become one of Mike's caretakers in January 2020. Many people think that the job is just playing with a tiger all day. Although Mike is very cute, we never 'played' with him. There was always a barrier between us for our safety. The daily responsibilities of being a caretaker consisted of cleaning the night house, feeding Mike, providing enrichment, communicating with Facility Services about his habitat and with Dr. Baker regarding everything having to do with Mike, and sharing information with the general public about tigers and wildlife conservation in general. I also had to learn some new skills when I started this job. For example, making his food into 'art' for football games or improving my photography skills for his social media. It was a job with unusual hours and a huge amount of responsibility, but it definitely had its perks.

"Getting to work with and learn about Mike VII was my favorite part of the job. I learned how intelligent he is and how he can remember people quite easily. It only took him a few days before he was able to recognize my face and would run up to greet me. He even recognized my car when I drove by the habitat. I was also able to see how curious and playful he is. He is always very aware of his surroundings and can tell if there's something different. For example, it took him a few days to get used to us wearing masks during the COVID-19 pandemic. We wore masks because tigers are susceptible to COVID-19. Since this was the first time Mike VII had seen anyone wearing a mask, he was very confused by the whole situation. He walked into the night house, and as soon as he saw us wearing masks, he turned around and went right back outside! He eventually came back inside and adapted to our altered appearance fairly easily. He was still able to recognize us, and we were still able to perform our responsibilities while keeping him safe.

"Another time he is aware something is different is on Sundays when he gets an oxtail as a treat. He always seemed to know when it was Sunday because he was more active and chuffed at me more than usual. Oxtails have always been his favorite, so he got really excited whenever he saw me open the oxtail box. Being a caretaker for Mike VII was a wonderful and unique experience for me. I saw him nearly every day (and sometimes twice a day) for two years, so he became a huge part of my life. And even though I miss him dearly, I was very excited to pass the torch to the new caretakers, so they can learn and bond with him as well." *Dr. Schneider works as a small-animal veterinarian in private practice.*

Erin Guillory (2022–present): "One of the moments that sticks out the most to me during my training to become Mike VII's caretaker was the very first time we went to the night house. Amelia and I had just been selected for the role, so our excitement was high, but it wasn't until that very first moment when Mike walked into the night house that it hit me. Here was this magnificent, intelligent animal, and I have the honor of working with him and caring for him for the next two years. That moment of impact will stay with me forever. Since that day, I have gained a new respect for Mike VII. His personality shines through his actions; he can be quite goofy at times and is very playful. It's amazing to see how powerful he is when he swats and lunges at his toy balls.

"It is also very special to see the unique bond that Mike has developed with each of his current caretakers. He interacts with them differently, which I find very interesting. Our training period has been a lot of work, but I can honestly say that there hasn't been a single day that I've woken up and not been excited to go. This job is incredibly rewarding, and I am still so grateful to have been selected for this opportunity. Working with a tiger is such a unique experience, and it has been fun to start developing my own bond with Mike VII. Learning his quirks and personality has been my favorite part of the job because it allows us to take better care of him, especially when it comes to knowing what his preferred type of enrichment is. Being selected to serve the LSU community in such a unique way is something I will treasure forever. Throughout the application process, thoughts of 'can I really do this?' sometimes ran through my mind; now that I have the position, I couldn't imagine it any other way. Mike VII is an amazing creature, and working with him is incredible." *Erin is a student in the LSU School of Veterinary Medicine.*

Amelia Talbot (2022–present): "I was a student at LSU when Mike VII arrived, and I went to his habitat frequently. Many of my memories at the tiger habitat are filled with laughter, as Mike VII is very playful. Even though I thought I had a good grasp on his personality, it was not until my first week in training as a caretaker that I began to fully appreciate how dynamic his personality really is. It has been interesting observing how the two caretakers at the time [Cora and Katie] interacted with Mike VII. Both of their relationships with him were strong but different. I am excited to take everything Cora and Katie taught me and form a unique relationship and bond with Mike VII.

"I was amazed at how intelligent Mike VII is and how quickly he came to recognize us. After one week, he responded to the sound of my keys and voice. I was outside the habitat, on the opposite side from Mike, and he came running across to me at the sound of my keys jingling. It is an indescribable feeling, having a tiger excited to see you." *Amelia is a student in the LSU School of Veterinary Medicine.*

Katie Miley with Mike VII's new caretakers, Amelia Talbot and Erin Guillory. (Katie Miley)

CHAPTER 2

BASICS ABOUT MIKE

Day-to-Day Life

Mike follows a daily routine that doesn't vary much throughout the year. Each morning, one or both of Mike's veterinary student caretakers arrive at the habitat and walk the entire yard to confirm that the fencing and locks have not been tampered with in the night, and that nobody has thrown harmful objects into the yard. When the students verify that the yard is safe and secure, they release Mike for the day. Normally, Mike has no access to his night house during the day. He stays outside playing in his pool or stream, watching people, or playing with his boomer ball—but most of all sleeping, as is normal for cats of all species.

If the outside temperature drops below 40°F and it is either raining or snowing, Mike is given access to his night house. Even when the weather is bad, however, he usually chooses to remain outside. Mike gets in his pool in all kinds of weather, including when it's snowing. The hairs that make up his undercoat are so dense that the water probably doesn't contact much of his skin, allowing him to enjoy his pool even in very cold weather.

Mike is only fed at night and inside his night house. This practice encourages him to come in each evening. It is important that Mike spend the night inside his night house so that his caretakers can examine him and to protect him from any human mischief. Occasionally, Mike won't come inside because he is not hungry or because something interesting is going on outside, like a track meet at the Bernie Moore Track Stadium or a basketball event at the Pete Maravich Assembly Center next to his habitat. On the

left Mike VI in the snow. (Christine Russell)

opposite Mike VII takes a much-needed nap from the rigors of serving as most awesome mascot. (Cora Schneider)

nights Mike won't go inside, his student caretakers notify his veterinarian, as well as the LSU Police, who make extra rounds in the night. Ordinarily, however, Mike is happy to come inside each evening, where he is fed and where his bed of pine shavings waits for him.

On Thursdays, the LSU landscaping crew freshens up Mike's yard. This keeps the yard looking nice and ensures a healthy environment for Mike. For many years, in January, some of Mike's friends have donated their Christmas trees. These trees are placed in the enclosure. Mike loves to rub against them because the sap smells so good! Of course, he also enjoys "attacking" them and dragging them around his yard.

Veterinary Care

Qualified veterinary care is an extremely important requirement for keeping a live tiger. And providing veterinary care for a 400-pound apex predator

Dr. Bivin visits Mike IV.
(Prather Warren, University Relations)

poses significant challenges. The veterinarian can't bring the animal out of his enclosure for a physical examination, blood or other sample collections, or close observation. While Mike could be sedated or anesthetized, these procedures bear great risk and so must only be done when they are essential. LSU is fortunate to have an excellent School of Veterinary Medicine. However, that hasn't always been the case. In 1955, when the aged Mike I became ill, Dr. Joe Dixon of the LSU Department of Veterinary Science was asked to provide care for the tiger. This was a tall challenge, since there was then no veterinary school at LSU! Even so, this was a significant advance in the overall care of the LSU live tiger mascots. From that point forward, Dr. Dixon or an associate from the Department of Veterinary Science regularly attended to the medical needs of the tiger, although overall responsibility for Mike remained with the Athletic Department.

In 1976 Dr. W. Sheldon Bivin of the LSU School of Veterinary Medicine took over as veterinarian for Mike III. As head of both the Division of Laboratory Animal Medicine and the exotic animal medicine service in the School of Veterinary Medicine, Dr. Bivin was the natural choice to become Mike's next veterinarian. This occasion also marked the formal transfer of responsibility for the tiger from the Athletic Department to a veterinarian.

TOOTH AND FOOT PROBLEMS

Dr. Bivin had to deal with several substantive medical issues with Mike IV. These medical challenges came from the most unexpected causes. Occasionally it was Mike that created a problem, while at other times trouble came from elsewhere. For example, there was the time when somebody turned a water moccasin loose in Mike's cage, and the time Tulane students attempted to paint him green. In March 1978, Mike IV broke a canine tooth while trying to get hold of a tire outside his cage and therefore needed a root canal procedure. Since veterinary dentistry was still in its infancy, Dr. Bivin recruited his personal dentist to perform the procedure. Mike IV loved to bite off pieces of tires and garden hose, chew them up, and swallow them. Twice, he consumed at least four feet of hose and required fairly intensive medical care for several days to help him pass all of the pieces. In addition to these episodes, he sometimes had wire cuts, intestinal problems, and sore feet. And since this roughly 500-pound patient had an "attitude," it was very difficult to provide medical care for him.

Mike V likewise presented his caretakers with numerous challenges. Perhaps his most common problem was sore feet caused by pacing on the concrete in his old enclosure. One of the most important renovations made to Mike's enclosure was the elimination of the concrete surface. The new enclosure allowed Mike to walk on grass—a huge improvement!

Like his predecessors, Mike V was plagued with dental problems. In April 1997, he somehow broke his upper right canine tooth, exposing the pulp cavity. Because a tiger's canine tooth is so long, special equipment is needed to extract the entire root. I contacted a veterinary dentist in Michigan who had the needed equipment and had performed the procedure on several tigers. He agreed to help Mike. I paced while Mike lay anesthetized on the operating table, waiting anxiously for the surgery to end.

With that procedure, Mike V became the third LSU tiger to require a root canal. Dental problems are common in tigers and are, indirectly, the leading cause of their death in the wild. As tigers age, their teeth become more brittle and are easily damaged or broken. Mike's team works hard to

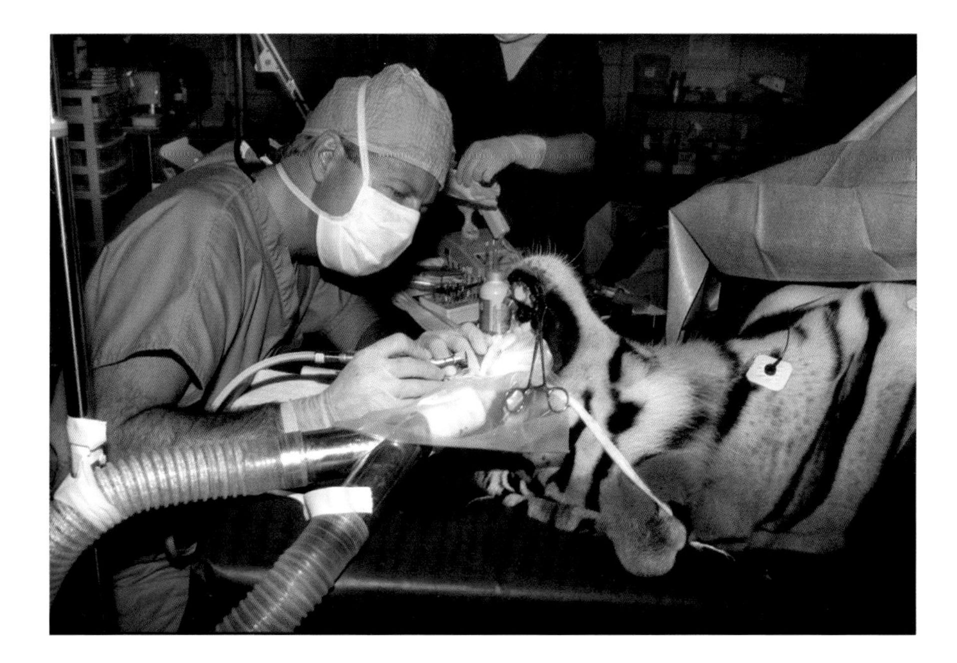

Dr. Alfred Stevens performs a root canal on Mike V. (Harry M. Cowgill)

above Mike VI gives a big yawn and shows off his beautiful teeth. (Christian Folk)

opposite Mike VII mauls his boomer ball on a snowy day in Baton Rouge. (Katie Miley)

preserve and maintain the health of his teeth. Mike is given a weekly oxtail in order to keep his teeth clean and to maintain his jaw strength.

On July 13, 2000, with the assistance of Dr. Gordon Pirie, I removed a tumor from the abdominal wall of Mike V. Mike is visually examined twice daily, and I had noticed the enlargement while observing Mike a few days earlier. An ultrasound-guided biopsy confirmed that it was a tumor. While Dr. Pirie and I were removing the tumor, Dr. Kem Singletary and Dr. Alfred Stevens, a local practitioner with special expertise in veterinary dentistry, performed dental repairs. Dr. Stevens had assisted with Mike's root canal in 1997 and now was equipped to do the work himself. The dental procedure was necessary because the unwilling patient had damaged a tooth during the ultrasound examination of the tumor.

Mike V had other medical conditions, in addition to foot problems and broken teeth. In November 2001, he developed a temporary thyroid problem, which resulted in considerable weight loss and hormonal changes. Fortunately, however, the problem passed, and Mike returned to his normal, healthy condition. As Mike's veterinarian, I had anticipated that medical problems would become more common as he aged. In that regard he was certainly like the rest of us. However, one can imagine the difficulty in providing excellent medical care for an aging animal that prefers not to be handled by his doctor.

Prior to the onset of cancer at ten years of age, Mike VI was remarkably healthy. Compared with wild tigers, captive tigers are generally very healthy since they are free from nearly all infectious diseases and do not have to capture and kill their food or defend their territory. In addition, Mike benefits from the highly qualified veterinarians and advanced clinical resources of LSU's excellent School of Veterinary Medicine.

Mike VII has been extremely healthy at LSU. The only significant medical issue he has experienced has been with feline idiopathic cystitis. However, now that the COVID-19 pandemic has largely passed and his caretakers are once again touching him (carefully) on a regular basis, even that condition has resolved.

WEIGHT CONTROL

Some guests notice that Mike seems to gain weight each summer, then slim down each winter. This is a natural weight fluctuation and likely occurs because Mike is not as active in the summer heat and so puts on a few pounds. In the winter, not only is Mike more active, but he burns calories staying warm, even though the winters in Louisiana are quite mild compared with those in the wild tiger's normal range. Weight control is an important aspect of Mike's care. It is common for captive animals to become too heavy, resulting in inactivity, joint disease, and other health problems. As his primary caretaker, I made sure that Mike remained at just the right weight for each season of the year.

Mike's Student Caretakers

LSU's first live tiger mascot lived at the City Park Zoo in Baton Rouge. In 1938, Mike I was moved onto the LSU campus, where he became the responsibility of the Athletic Department. For many years, Hickey Higginbotham had general oversight of Mike, while his day-to-day care was provided by members of the swimming and diving teams, hired by the Athletic Department. From the 1940s through the 1970s, the student caretaker position was passed down within several LSU fraternities.

The lives of the many "tiger keepers" are themselves interesting stories. At least one, Charles Fourrier, Mike's caretaker from 1938 to 1942, enlisted for combat duty in World War II. Fourrier lost his life in an aircraft landing accident at Pearl Harbor, thereby giving his life in service to his country. Another former caretaker nearly shared the same fate. Osce Jones became a P-51 fighter pilot. He was shot down over France and spent time in a German POW camp.

When daily care of Mike the Tiger transferred to Dr. Bivin at the LSU School of Veterinary Medicine, he began to recruit student caretakers from among the veterinary students. Every two years Mike's veterinary student caretakers change, requiring Mike to form relationships with two new

Caretakers Nick Fleissner (*left*) and Tré Gildersleeve visit with Mike VII. (Ginger Guttner)

Tré Gildersleeve with Mike VII. (Nick Fleissner)

caretakers. While the tigers eventually adjust to this change, the departure of their student caretakers, with whom they have become familiar, can be very hard on them. For example, following the 2008 departure of Mike VI's caretakers, he lost seventy pounds! He clearly missed them. But, as Mike VI developed relationships with their replacements, his weight returned to normal. As he matured, each subsequent turnover of students seemed to have less impact on him.

Mike's relationships with his student caretakers are not all the same. Each tiger seems to develop a "favorite" student caretaker. One caretaker recalled that Mike VII would not come inside his night house one night until her teammate played her voice over a cell phone!

Only a few of Mike's veterinary student caretakers have gone on to practice exotic animal medicine. The remainder are spread across a wide range of sectors within veterinary medicine. I am sometimes asked what I most enjoyed about serving as veterinarian for Mike the Tiger, and the answer is always the same. While I certainly appreciated working with Mike, my favorite part of the job was mentoring the veterinary student tiger caretakers. I saw them mature into even more competent professionals through the responsibility of caring for Mike. It has been a pleasure to play a small role in their professional development.

Mike on Social Media

Ginger Guttner, communications manager for the LSU School of Veterinary Medicine, serves as the spokesperson for LSU's live mascot. That is, she handles media requests and manages Mike's social media accounts. In the fall of 2009, Ginger discovered that someone had created a Facebook page for Mike VI. After some investigation, she found that the page had not been created by anyone at LSU with the authority to do so. When LSU informed Facebook that the page was not authorized, they took the page down but wouldn't reveal its owner. Within a day or so of that incident, LSU created an official Facebook page for Mike VI.

In the beginning, Ginger wasn't sure how to best run the page. "Mike is an amazing animal, but he doesn't do a whole lot," said Ginger. It was a struggle to know what to post and how often. In addition, Ginger was concerned about posting in first person from Mike's point of view. She was afraid that the fans would complain and not accept it. And she questioned whether it was right for her to presume to speak as Mike the Tiger.

During the first nine months of Mike's official Facebook page, Ginger posted once or twice a week from her own point of view and used the book I wrote with Dr. Bivin, *The Roar of LSU,* for content. She basically used it as a "Did You Know?" space for facts about Mike VI and the previous tigers. There was little engagement with fans. In August 2010, everything changed when a father posted a precious photo of his then-two-year-old son with Mike in the background at the habitat. The boy was five at the time of the post but was in the hospital, and his father was asking for prayers. Ginger first thought of posting in third person as usual, but realized that this post was something the father might show his son. She thought, "If this kid writes a letter to Santa Claus, he doesn't want to hear back from Santa's secretary." So she posted as Mike. No one complained. No one fussed. Some people complimented Mike for being sweet.

From this point on, Ginger started posting as Mike, still using our book for content. Then one day, with little time to spare, she posted something silly. The photo that day showed a side view of Mike, and Ginger posted, "Do these stripes make me look fat?" This comment was not intended to be educational or even interesting. However, it sparked a conversation about Mike's weight and his diet. Something not intended to be consequential ended up educating and engaging people. Ginger then changed the tone of Mike's

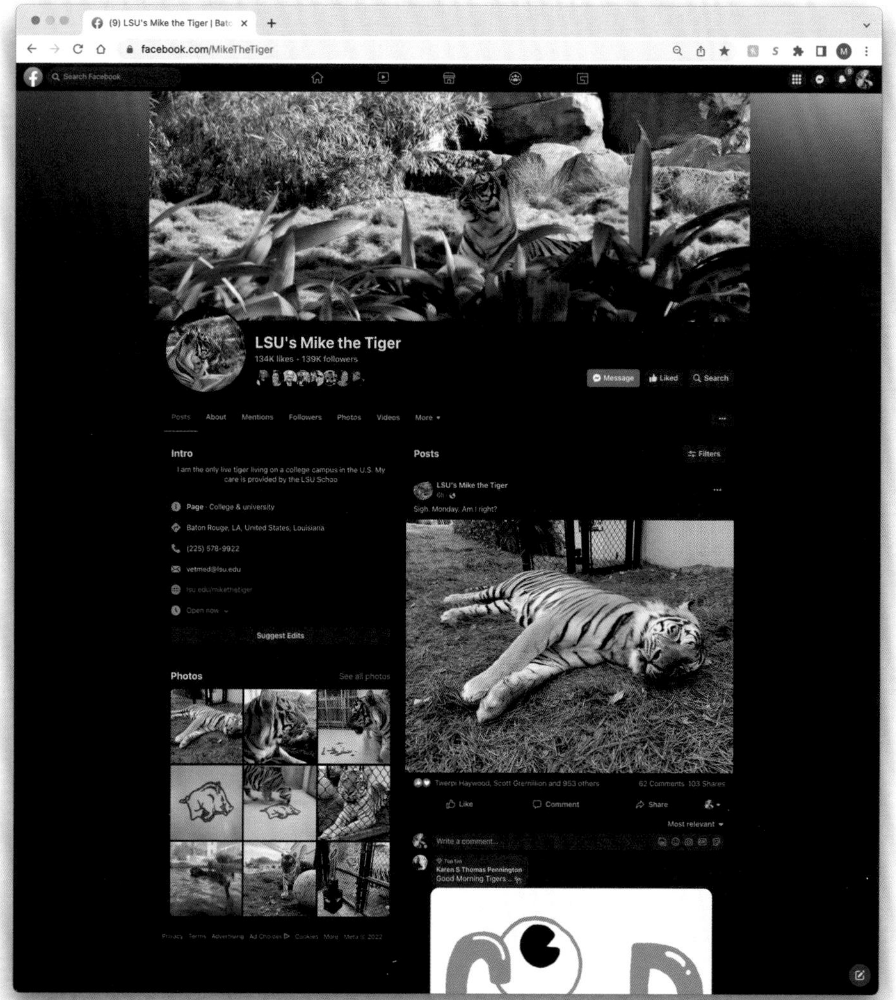

Screenshot of Mike's Facebook page.

Facebook page. It became more fun, though still occasionally educational. A Twitter account was added and later Instagram.

In May 2016, when Mike VI was diagnosed with cancer, he had approximately 100,000 followers on his combined social media platforms. The announcement of Mike's illness doubled his number of followers in a matter of months. Social media was vital in keeping fans up-to-date on his treatment and prognosis. Before the diagnosis, Ginger typically posted for Mike five times a week on weekdays. She posted on weekends of LSU home football games and, if a holiday fell on a weekend, to let people know when Mike was outside. LSU is closed for holidays, but the habitat is always accessible for visitors. After the cancer diagnosis, she started posting daily to show fans that Mike was okay and in his yard as usual. She also used social media to share information about his treatment and his medical team. Many fans had questions and comments, both public posts and private messages, and Ginger monitored and replied to them.

In 2017, Ginger used both social and traditional media to announce the arrival of Mike VII. She resumed posting five days a week until the COVID pandemic began, when she returned to daily posts. With LSU closed, Ginger wanted to let people know that they could still visit Mike (though with masks and behind an extra barricade), and she wanted to show his fans that Mike was fine. This practice will continue since people appreciate it so much. By April 2022, Mike's followers totaled almost 300,000.

Every morning, Mike's veterinary student caretakers text photos to Ginger to let her know that he's outside. Ginger uses these in her posts, which often joke about Mike's good looks or the fact that he seems to like swimming and sleeping more than anything else. She checks his accounts regularly and responds to messages as needed. Any questions that Ginger can't answer came to me. If a question is posed directly to Mike (e.g., "Hey, Mike, how much do you weigh now?"), Ginger responds as Mike (e.g., "I weigh about 410 pounds."). If the question is directed to the administrator (e.g., "How much does Mike weigh?"), Ginger responds accordingly ("Mike weighs about 410 pounds."). Often fans will step in and reply to questions before Ginger gets a chance to do it, but "Mike" will then confirm or correct the reply if necessary.

Social media is a useful tool, allowing Mike's fans to check in on him every day. They learn about the mascot program and his daily care. Ginger can also address rumors about Mike and promote global tiger conservation. Because of the success and impact of Mike's social media presence, Ginger has been approached by other universities asking for advice on using social media for their mascot programs. There is a close community of university employees who work with live mascots around the country. As mascots from other universities pass away, Mike expresses condolences, and when they retire, he wishes them well. "Dr. Baker always says that Mike represents all that is good about LSU," said Ginger. "Social media allows us to share that goodness with hundreds of thousands of people every day. It is one of my favorite things about my job, and getting to see a live tiger up close never ceases to thrill me."

Meat art for a game against Auburn. (Christian Folk)

MEAT ART

One of the most popular types of posts on Mike's social media is "meat art." This began in 2014 when caretaker Macy Trosclair sent Ginger a photo of Mike's food molded into the shape of the University of Alabama's cursive "A" logo. This was just something Macy did as a joke. After Ginger showed the photo to Lindsey Clemones and Trevor Davis, Mike's caretakers from 2014 to 2016, Trevor asked if he could mold Mike's dinner into the opposing team's logo before every football game. I agreed, and "meat art" was born on social media.

The student caretakers use Mike's regular food, which has the consistency of ground meat, to create the emblem of the opposing team's university, then allow Mike to eat it. The "meat art" is posted on Friday night before Saturday LSU football games. Not only is this another form of enrichment for Mike, but LSU fans love it! In fact, the practice has gotten more popular each year and has been featured on CNN, ESPN, CBS Sports, and Fox Sports. Many people have asked if the students use some type of aid in creating the designs. Some are created with the aid of a projector, while others are created freehand. Katie Miley recalls that in 2020, even with the help of a projector, she and her teammate took three hours to complete the University of Florida alligator, but LSU won the game, so it was worth it!

Interactions with Mike

One of the qualities we look for and cultivate in Mike the Tiger is positive interaction with people. It is important that Mike be part of the LSU community, rather than hide in the back of his yard unwilling to engage with visitors. Since he is a wild animal, there are limits to what we can expect in animal-human interactions, but there is benefit to having a tiger that is inquisitive, confident, and interested in what is going on around him.

MIKE I

Mike I was not overly fond of people, as several of his caretakers attested. Eddy Contine, one of his earliest student caretakers, once found himself face-to-face with Mike I. Contine had forgotten to close a gate before entering the enclosure. He was so frightened all he could stutter was: "Hic . . . Hic . . . Hic . . . y!" Hickey Higginbotham grabbed a pan and started running it along the caging while calling out "Sheik!"—Mike's former name. The noise distracted Mike long enough for the terrified student to make his escape.

Mike I's next student caretaker, Charles Fourrier, also once forgot to close the gate separating himself from Mike. Fortunately for all, he managed to get out of the cage before Mike could attack. A similar incident occurred with caretaker Osce Jones. A few years later, Brittain Briggs once got too close to Mike's cage while chipping ice off the lock. Seeing an opportunity, Mike reached a paw through the bars and slapped Briggs to the ground! Jere Melilli's brush with Mike was even more dangerous. He was inside Mike's house washing down the floor with a hose when his foot slipped toward the cage holding Mike. In an instant, Mike reached through the bars and slashed Melilli's shoe. There is no doubt that Mike was trying to get the student's foot inside the cage with him.

Handler Hickey Higginbotham plays with Mike I, ca. 1937. (Leon Trice, *Baton Rouge Advocate*)

MIKE II

Both tigers that served as Mike II appear to have been relatively calm and easy to work with. That might have changed as they reached maturity. However, since both died prior to that point, there is no way to know how they would have behaved as adult tigers.

MIKE III

It was apparent to his keepers early on that Mike III was noticeably calm. His calm nature may have saved the lives of at least four of his caretakers. The first was Alvin Meyer Jr., Mike's student caretaker from 1961 to 1963. One day Meyer forgot to make sure that Mike was safely locked in his tiger house before entering the enclosure. While going about his business, he came face-to-face with the big cat. Mike had wandered out of his tiger house looking for his dinner. Fortunately for all, he was more interested in his supper than the student. Needless to say, Meyer beat a hasty retreat.

Meyer had another brush with disaster when Mike III grabbed his foot with his paw. Meyer had been sitting in the tiger house with his feet propped up against the cage bars when Mike decided to play with Meyer's leg. After that, Meyer decided he'd better keep his feet a bit farther from the cage. Next, Joel Samuels reports that on at least three occasions he accidentally entered the enclosure thinking that Mike was locked up when he wasn't. During two narrow escapes, Mike chased the fleeing student, jumping up on the cage door just after Samuels closed it behind him. Bob Lank Jr., Mike's caretaker from 1968 to 1972, develops sweaty palms and a quickening pulse when he recalls the time he too found himself in the cage with Mike III. Lank was cleaning up in the outer enclosure when a bystander asked why Mike didn't attack him. He confidently replied, "Because of that gate." When the woman asked, "What gate?" he turned to find that he had forgotten to

Mike III in 1964.

lower the gate that locked Mike inside his tiger house. Fortunately, Mike was calmly eating his dinner and didn't notice the frightened student as he dashed out of the enclosure to safety.

Philip Aman, Mike III's caretaker in the spring of 1972, recalls a similar experience. He had entered Mike's enclosure to clean up and forgot to lock the tiger inside his tiger house while Mike ate his supper. Aman was hosing down the enclosure when he noticed his potentially fatal error. He realized that he had to make his escape before Mike noticed him. The only problem was having to cross in front of the open gate of Mike's tiger house. Having to choose between a risky escape and certain danger once Mike finished his dinner, Aman made a dash for the outer gate. As he rushed by the open door, he startled Mike, who turned in time to see the student reach safety. After composing himself, Aman dropped the gate, locking Mike in the house, and resumed his duties inside the enclosure.

MIKE IV

Mike IV was the largest and strongest of the LSU tigers, though all have been capable of severely injuring a person. There is a report describing how 6'9", 295-pound LSU defensive end Leonard Marshall, later of the New York Giants, once had a tug-of-war match with Mike IV over the hanging tire in his enclosure. According to eyewitness accounts, Mike pulled down on the tire with such force that Marshall, who was hanging on the other end of the rope trying to lift the tire away from Mike, was himself lifted into the air. The watching football team was amazed at Mike's incredible strength. Unless you have seen it for yourself, it is difficult to comprehend the strength and speed of tigers. Mike demonstrated to all why the tiger is the "King of the Jungle."

MIKE V

Mike V was hand raised, and many people had the opportunity to touch and pat him as a cub. Direct human contact continued until he reached 250 pounds. Rollie Norris was the last student caretaker to have contact with Mike V with no bars between them. One day Norris took a photographer into Mike's enclosure, thinking that Mike was safely locked on the other side of a gate. Much to their surprise, however, the gate was open. Mike came through the gate and ran right over the terrified photographer. When he came back around for another "attack," Norris grabbed the photographer, threw him out of the enclosure, pushed Mike firmly on the nose, and got himself

Mike V was hand raised and accustomed to being handled by people. (Jim Zeitz, University Relations)

out before Mike decided to get aggressive. While it is unlikely that Mike would have attacked with the intent to injure, one can't take chances with a 250-pound tiger.

MIKE VI

Mike VI was the most interactive and affectionate of all the Mikes, and he remained so throughout his adult life. Upon his release into his yard for the first time, many people were surprised by his large size and his energy level. Photos and videos of the new tiger playing and interacting with visitors through the glass of his enclosure spread quickly over the Internet. Through

Mike VI interacts with his fans. (Michael A. Westphal)

these images as well as personal visits, the LSU community developed an intense love for Mike VI.

Developmentally, Mike VI was roughly equivalent to a ten-year-old boy at the time he came to LSU. Can you imagine how much had changed in his life? Among other things, he had left his original family for a new one without other tigers. With all of the changes in his life, Mike needed companionship, and he found it in his caretakers, with whom he quickly formed strong relationships. Tigers normally live alone in the wild as adults, but they enjoy the companionship of other tigers while they are young. For Mike, humans would have to do. He easily came to recognize his caretakers. For example, Mike learned the fragrance worn by caretaker Kelly Folse, and as soon as he caught the scent of her perfume, he would perk up and look for her.

MIKE VII

Like all of LSU's previous tigers, Mike VII has a unique personality. He is inquisitive and very interested in people, particularly small children. Even though he has reached maturity, he is still very playful. As noted earlier, tigers come to recognize people whom they regularly see. Mike VII is no different. Interestingly, he has a positive reaction to many of the people who frequently visit him, and a somewhat negative reaction to others. There is no way to know why he "likes" some visitors more than others, since none of them have direct contact with him.

I visit with Mike VII.
(R. C. Boccaccio)

Mike VII is more easily agitated and more aggressive than the two tigers before him. In that way, he is more like Mike I and Mike IV, though thankfully not that aggressive. Also, he doesn't like surprises, so his caretakers make extra effort to keep his routine consistent. Fortunately, Mike VII seems very comfortable with his caretakers. I was amazed at how quickly he came to recognize me—within twenty-four hours of his arriving at LSU. In comparison, it took Mike VI several days before he seemed to recognize and trust me.

It is important that Mike like and trust his caretakers so that he will approach them. This allows them to look him over carefully, and it is a sign of his psychological well-being. Despite Mike's apparent affection for his caretakers, they can never let down their guard around him.

Mike's Diet

Mike's diet has always been a critical component of his care. Unfortunately, knowledge about the nutritional requirements of large exotic cats had not developed very far in the days of the first three tiger mascots. Mike I ate a hearty daily meal of 8–12 pounds of meat, sometimes covered with cod-liver oil, a "slab" of salt, and regularly, as a special treat, three eggs and a quart of milk to maintain his sleek coat. While this might sound like a great meal for a carnivore, the heavy dependence on meat resulted in a diet imbalanced in its calcium-to-phosphorus ratio. Such a diet often leads to the development of soft-tissue mineralization and arthritis. Both tigers named Mike II were fed a similar diet. The first Mike II didn't live long enough to develop nutritional problems. However, the bone fractures experienced by the second Mike II were very likely associated with an imbalanced diet.

Mike III lived a long and healthy life on about 12 pounds of horsemeat daily plus cod-liver oil, vitamins, and minerals. This mascot was also regularly given freshly killed chickens, which his veterinarian had heard were good for tigers. The addition of whole chickens was an improvement, since it provided a bit more than just muscle meat.

Beginning with Mike IV, LSU's tigers have been fed a commercial carnivore diet. The diet comes in 5-pound logs and looks and feels like hamburger. But it is more than that. It contains beef, fish, vegetables, and other ingredients. It is a balanced diet intended to replicate the tiger's natural food sources—that is, whole animals.

On Thursdays, when the LSU landscaping crew works in Mike's yard, he is held inside his night house and, as a treat, given a frozen oxtail. Occasionally, his caretakers also acquire organ-meat treats for him, including pig or cattle hearts or kidneys, and sometimes even whole chickens or turkeys, as on Thanksgiving. On Mike's birthday and other special occasions, his student caretakers make him a "cake" out of his food, topped with a few strips of bacon and frozen goat's milk "candles" indicating his age.

Sometimes Mike receives frozen "bloodsicles." These are made from blood obtained from healthy animals. Mike is usually offered pumpkins around Thanksgiving and watermelon in the summer. He doesn't consume much of the fruit, but he sure enjoys tearing it up! Mike is never provided with live prey. His caretakers do not want him in the habit of killing things, both for the sake of the prey animal and as a safety precaution in the event of an escape.

Mike VI enjoys a frozen treat of tiger food, goat's milk, and an oxtail. (Christian Folk)

Mike's Habitat

Mike's habitat is an essential aspect of his care, as this is the place where he spends his life. The structure of Mike's enclosure has changed greatly over the years. As mentioned earlier, Mike I was housed at the City Park Zoo on the north edge of the LSU campus, overlooking the LSU lakes. Mike I moved down the hill onto the LSU campus after he got his leg and paw between the bars of his cage and slashed his caretaker, Hickey Higginbotham. A tiger house and small outdoor cage were built next to the football stadium in 1937, at a cost of about $3,500. Mike's new home was dedicated by Governor Richard W. Leche prior to an LSU-Alabama baseball game that year.

The 1937 enclosure remained Mike's home for many decades. In 1981, it was enlarged while Mike IV was temporarily relocated to the Little Rock Zoo. When the expansion was completed, it took Mike some time to adjust to his higher style of living. Byron Garrity, his caretaker at the time, described the tiger's first few weeks in the refurbished surroundings. Mike was very wary of his new swimming pool. After about a week, however, he came to love it. Mike also didn't know what to make of the grass that had been added to the enclosure. Being raised in captivity, he had never seen grass. He walked around it and put his paw on it, but he wouldn't walk on it for several days. Unfortunately, the grassy area had to be covered over during the reign of Mike V. His urine created dead areas in the grass, which quickly turned to mud. It was feared that so much mud might promote foot infections.

By 2000 it became evident that if LSU was to continue housing a live tiger on campus, a new habitat would have to be built so that Mike V and all future Mikes could enjoy a more healthful environment. University officials had discussed the need for a new home for Mike for several years. But now, the deteriorating condition of the enclosure, coupled with Mike's chronic foot problems associated with prolonged housing on concrete, required action. A larger, more environmentally enriched facility with water, trees, shrubs, grass, and other natural substrates was envisioned.

The Tiger Athletic Foundation took on the monumental project and raised the $3.2 million needed for construction and endowment. The fundraising drive kicked off in May 2002 at the home of head LSU football coach Nick Saban. The response from the LSU community was tremendous. In 2003, local artist and avid LSU fan George Rodrigue, famous for his "Blue Dog" paintings, contributed his prodigious talents to the cause. The artist painted a swamp scene of Mike the Tiger in a cypress tree that was an immediate hit. Print sales raised more than $1 million toward a new home for Mike.

Mike's new 13,000-square-foot enclosure was designed by Ace Torre, an LSU alumnus who had created zoo animal exhibits all over the country. The

Mike V lies low in his pool. (Harry M. Cowgill)

design incorporated an Italianate campanile as an exhibit backdrop as well as an LSU-style colonnade façade for viewing. The habitat included a waterfall, stream, and pool, and allowed Mike to get among the shrubbery and out of the weather at all times of the year. Before demolition of the original enclosure, the frescoes were removed from the old night house and preserved as historical artifacts.

Frescoes from the original enclosure and construction of the new enclosure, 2005. (Berkman Manuel of Gibbs Construction)

After a nine-month stay at the Baton Rouge Zoo during the construction of his new habitat, Mike V returned to LSU on August 26, 2005, just three days before Hurricane Katrina slammed the Gulf Coast. Mike rode out the hurricane in the safety of his new night house. For several days he watched with great interest as helicopters ferried New Orleanians to the adjacent Pete Maravich Assembly Center, where a medical triage unit had been established. After things settled back down, and once he had assured himself that he was the only tiger around, Mike adapted quickly to his new enclosure. As hoped, the larger, environmentally enriched space was much more interesting for Mike, providing wholesome surroundings for him and for many tiger mascots to come.

Every part of Mike's habitat must promote his mental and physical health and well-being. The night house must be kept clean and be organized in ways that facilitate Mike's movement and care. There can't be any structural components, including sharp edges, that could injure him. The temperature, humidity, lighting, and air quality must support his health and comfort. He must have surfaces on which to walk, stand, and lie down that are smooth enough to be comfortable and cleanable, but rough enough to prevent slipping, which could injure a hip or leg.

left Mike V in his spacious new habitat. (Dawn Forste)

right Mike VII cools off in his pool. (Katie Miley)

Likewise, the habitat yard must be sufficiently spacious to permit Mike to move about freely. Many years ago, zookeepers were concerned about the amount of pacing done by some captive tigers. However, as biologists began fitting wild tigers with radio collars, they discovered that the males were walking several dozen miles per day, patrolling their territory—which itself could be up to several hundred square miles, depending on the availability of food and mates. So it can be a very normal thing for a tiger, particularly a male tiger, to walk a lot in a day. Mike's yard provides him with ample space to "patrol."

Water is a vital component of a tiger's habitat. Tigers are natural swimmers and are rarely found in the wild away from bodies of water. Mikes V, VI, and VII have enjoyed the stream and pool in their enclosure. The habitat yard is purposely asymmetrical, giving a unique view from any location, both for guests looking into the yard and for Mike looking out of the habitat. This creates interest and is another form of environmental enrichment. The trees and other foliage provide shade and beautify the habitat. The hot/cold rock platform provides a temperature-controlled place for Mike in extreme weather conditions. The glass windows offer unobstructed views into and out of the habitat, allowing guests to see Mike clearly and Mike to look out of his

TIGER STADIUM

yard from any angle. Finally, the system of gates and mesh surrounding the habitat are intended to keep Mike secure from irresponsible people, and to keep people safe from a 400-pound carnivore!

SANCTUARY STATUS

The US Department of Agriculture (USDA) inspects LSU's tiger habitat and tiger mascot program annually to ensure compliance with the federal Animal Welfare Act. No significant issues of noncompliance have been identified in the past several decades. Because of this positive record of care, and the evolution of the mascot program to a sanctuary program, in 2017 LSU investigated accreditation with the Global Federation of Animal Sanctuaries (GFAS). However, despite LSU meeting all Animal Welfare Act requirements and meeting or surpassing all GFAS standards of care, the GFAS chose not to accredit LSU's tiger habitat because LSU scientists conduct biomedical research using animals.

It is important to note that no research is performed on Mike the Tiger. In addition, LSU's biomedical research involving animals is conducted using humane methods and is overseen and regulated by the USDA, the US Public Health Service Office of Laboratory Animal Welfare, and the LSU Institutional Animal Care and Use Committee. Finally, LSU's research animal program is fully accredited by AAALAC Int., a nonprofit organization that inspects, reviews, and accredits animal programs to ensure compliance with government standards of humane animal care. So the GFAS decision not to accredit LSU was purely ideological and had nothing to do with tiger care. In response to the GFAS refusal to accredit LSU's tiger habitat, the university decided to consider the habitat and tiger care program to be self-accredited.

ENVIRONMENTAL ENRICHMENT

I have referred several times to environmental enrichment, which is an important aspect of Mike's care. A bored, unstimulated tiger may develop harmful, compulsive behaviors. Mike's caretakers regularly devise new ways to keep him stimulated. For example, they may hide his food in different places, construct artificial prey for him to tear apart, or place unusual scents in his enclosure. Matthew Wheelock recalls purchasing a kiddie pool for Mike V. The plan was to get him used to the pool and then add fish. Surely Mike would love to watch fish swim. Unfortunately, the plan broke down before the addition of fish. It seemed Mike didn't need the fish; he was happy

opposite Mike VII surveys his habitat. (Katie Miley)

below A boiled egg hidden in plain sight as a treat for Mike VII. (Katie Miley)

right Mike VII eats watermelon. (Katie Miley)

lower right Mike VI with his boomer ball in the night house. (Christian Folk)

lower left Mike VII attacks a pumpkin. (Christian Folk)

Mike VII doing target training. (Katie Miley)

just to destroy the pool! Mike's caretakers tried pools of at least four different colors in hopes of finding one that he wouldn't want to destroy. They never found a suitable pool, so the plan was abandoned. The experiment was not a total failure, however, because Mike certainly was entertained in the process.

Other forms of environmental enrichment, in addition to the general construction and layout of the habitat, include personal contact with the caretakers, a boomer ball, Christmas trees in January, and diet variations such as bloodsicles, frozen goat's milk, bacon, pumpkins, watermelon, and an annual Thanksgiving turkey. Mike also receives "target training," which consists of clicking a small plastic device while holding a red plastic target up against the caging. Mike is expected to stand up and touch his nose to the caging at the point of the target, to allow his caretakers to closely examine his body. If he cooperates, he receives a food treat. The most recent tigers have differed in their willingness to cooperate.

MIKE

CHAPTER 3

TRAVELING WITH MIKE

The early tiger mascots traveled extensively. Mike I attended football games as far away as Birmingham, Alabama, to help cheer the team on to victory. Sometimes he was transported by train, but usually he traveled in his $300 trailer, donated by the Louisiana Highway Department. During Mike's first few years as mascot, he traveled around the state as a sort of publicity agent for the football team. He also occasionally accompanied the swim team as they gave exhibitions around Louisiana. It seems that Mike I even did some late-night traveling within Baton Rouge, courtesy of his student caretakers. Gordon Doré tells of loading Mike up late one night and taking him into town. Reportedly, onlookers were given the opportunity to see Mike in the trailer; then Mike was returned to campus and quickly unloaded. Revisiting the same parts of town, but this time with the trailer empty and the gate swinging open, brought about an entirely different response. People were convinced that the tiger was loose in their neighborhood and went scurrying for the safety of their homes. The students encouraged the deception by asking people if they'd seen a tiger. It is alleged that parts of the city closed up early that night!

During World War II, a plan was devised to transport Mike to the Audubon Zoo in New Orleans so that he would have access to adequate food, specifically, meat. LSU students did not like the idea of seeing their beloved mascot housed in Tulane territory. Some students even forced the student body president and vice-president to ride around campus on parade in Mike's trailer, pulled by seven male students, prior to a meeting with university offi-

opposite Mike I in his trailer with homecoming trophies displayed on top, 1942. (Courtesy Mrs. Osce R. Jones)

The original trailer during Mike II's reign, with cheerleaders riding on the top, 1956. (*Gumbo*, 1957)

cials to discuss the transfer. With so much opposition from the student body, the administration scrapped plans to ship Mike to New Orleans.

When LSU traveled to the 1947 Cotton Bowl, caretaker Hickey Higginbotham hitched Mike's cage to his car and headed for Dallas. Leaving Baton Rouge, it was 85°F and sunny. When they reached Shreveport, it was 15°F and sleeting! Not wanting to expose Mike to more inclement weather, Higginbotham requested and was granted permission to shelter Mike in a warehouse for a few days. Unfortunately, no one told the night watchman about Mike's arrival. When he entered the warehouse on his nightly rounds and shined his flashlight on Mike, the tiger let out a growl that nearly scared the man to death! It is rumored that the poor man never completely recovered from the experience.

Missing Tiger

December 1, 1950, marks one of most famous episodes in the history of the tiger mascots. The most believable account of the incident comes from Tom Freeman, who frequently assisted Mike's caretaker, David Melilli. On that day, Mike I was taken to New Orleans for the LSU-Tulane football game. It was customary to bring Mike to New Orleans the day before a game to avoid the traffic. Mike would frequently stay at Ye Olde College Inn under the eye of a night watchman. On this occasion, the watchman went to the rear of the inn to find a tarpaulin with which to cover the trailer. Melilli had gone to the Roosevelt Hotel. He returned to the inn about ten minutes later only to find Mike missing!

Melilli and Freeman drove around New Orleans listening for a tiger's roar. They didn't know that Mike had been taken to the Tulane football stadium and was being guarded by Tulane students and local police. The Tulane students had hatched a plan for their cheerleaders to present Mike back to the LSU cheerleaders at halftime. When Melilli and Freeman found Mike, they attempted to retrieve their precious mascot, but the police would not give him up. Finally, at literally the last hour, LSU officials demanded Mike's return and he was reluctantly handed over. The LSU students had just enough time to remove the blue and green crepe streamers draped on the trailer as Mike was entering the stadium. Unfortunately, there was no time to remove the green paint from the trailer, where the Tulane tigernappers had proudly announced the name of the offending university. This incident fueled the rivalry between the two schools for many years.

Trailer Tales

Over his years of traveling, Mike I survived at least two automobile accidents. The most serious accident occurred in 1951 on US Hwy 71, about fifteen miles north of Coushatta, Louisiana. Mike was returning from Shreveport in triumph after LSU beat Villanova 45–7. A motorist drove onto the highway shoulder and attempted to pass Melilli. Melilli was towing Mike's trailer with his car and was himself trying to pass a slower vehicle. Melilli swerved back into his lane, but in the process the trailer carrying Mike flipped over into a ditch. Fortunately, nobody, including Mike, was injured. A wrecker from Coushatta was called out and righted the trailer. After clearance from state troopers, the entourage was allowed on its way.

Mike I was unusual for a tiger in that he liked to stand up in his trailer while traveling. This was sometimes a problem, since he could spray urine right into the car towing him! That happened on at least two occasions during parades down Third Street in Baton Rouge. In both cases, a group of coed cheerleaders were sitting on the back seat of the towing car, a convertible, when all of a sudden they started screaming and diving for cover. Both Hickey Higginbotham and Jere Melilli would later laugh about it, but at the time they didn't dare!

Mike's new custom-built trailer, with cheerleaders on top, 1957.

In 1953, university officials decided that, due to his advancing age, Mike I's traveling days were over. Unfortunately, that policy appears not to have been followed completely. On November 27, 1954, Mike I was once again on his way to New Orleans for the LSU-Tulane game. A sudden stop by caretaker Bob Dodwell caused Mike's trailer to jack-knife and turn over on Airline Highway. Although Dodwell was slightly injured, Mike was thankfully unharmed.

During the course of his reign, Mike I eventually traveled to all of the schools in the Southeastern Conference. Occasionally, Mike and Higginbotham would get back to Baton Rouge after midnight. Rather than going to the trouble of returning Mike to his enclosure at such an hour, Higginbotham would park Mike's trailer inside his garage at home. The neighbors got used to hearing the sounds of a tiger coming from the Higginbotham garage!

Paul Marks Jr., caretaker for Mike III from 1963 to 1965, recalls a time when LSU was to play Ole Miss. It was rumored that the Ole Miss fans were planning a caper the night before the game. They planned to paint whatever they could find, including Mike III if possible, red and blue. Campus police encouraged Marks to spend the night guarding Mike's cage, but that didn't appeal to a student who liked his sleep. So Mike III was loaded into his trailer for a trip to the home of Marks's parents. There he spent a peaceful night in a closed garage, safe from any would-be Picassos. And the student got his good night's rest!

Because LSU competed in so many bowl games during the reign of Mike III, his caretakers have many stories to tell of their travels. Joel Samuels recalls taking Mike III to Dallas for the 1966 Cotton Bowl, when LSU was pitted against the Arkansas Razorbacks. Though a two-touchdown underdog, LSU won the game 14–7. Afterward, at the Cotton Bowl banquet, the governors of Louisiana, Arkansas, and Texas were deeply engrossed in conversation with one another when Samuels was introduced as "the Tiger Man." Suddenly all three governors stopped their conversation and turned their attention to him. Each was more interested in talking about Mike the Tiger than in talking to one another.

Samuels also recalls an LSU-Alabama game when Bear Bryant was the Alabama head coach. Samuels had a way of getting Mike to growl simply by saying, "Get 'em, Mike." As the Alabama players gathered around the tiger trailer, Samuels quietly gave Mike the signal. When Mike growled, it so unnerved the Alabama players that Coach Bryant took Samuels aside and gave him a stern lecture, advising him not to upset his players like that. As far as we know, none of the other tigers growled on cue.

For many years, the tiger trailer was usually pulled by a Ford Fairlane convertible provided by a local Ford dealership. Samuels was sometimes loaned the next year's Ford model to pull the trailer. These models were so new that many fans were seeing them for the first time. In 1968, an LSU fan donated a roadster painted in a tiger motif for use at the football games. Bob Lank Jr., Mike's caretaker at the time, reports that while the roadster was certainly beautiful, he was disappointed at losing the convertible. He had been

GR-RR
STP

anticipating using that car on the weekends when not pulling Mike. He says that the roadster was difficult to maneuver within the stadium and would frequently overheat. After the 1972 season, the roadster was retired, and tiger keepers began using trucks to pull the trailer.

opposite Tiger-striped roadster used to pull Mike III's trailer.

Mike III stopped traveling with the football team in 1972 when his trailer overturned on Highway 190 en route to a fundraising event in Shreveport. The accident occurred because an incorrectly sized ball had been installed on the towing vehicle. When the trailer hit a bump in the highway, it bounced clear of the towing vehicle, ran into the median divider at 45 mph, flipped twice, and collided with a tractor-trailer rig! Amazingly, Mike escaped with only a small wound on one paw. Traffic backed up for miles as various police units arrived on the scene. The trailer sustained so much damage that the door had to be cut off with a torch in order to free Mike. After that incident, it was decided that a police escort would accompany the tiger mascot whenever he left campus. And, to ensure Mike's well-being, off-campus travel was restricted. Those policies remain in force to this day.

Mike IV was quite a local traveler. He participated in Mardi Gras parades in Thibodaux, Baton Rouge, and New Orleans, and he accompanied LSU to the Sugar Bowl in 1985. Unfortunately, Mike's presence at the game was not enough to put the Tigers over the top, and LSU was defeated by Nebraska (28–10).

Mike V's first road trip was to the Superdome in December 1991 for a basketball game against Texas (which, with the help of Shaquille O'Neal, the Tigers won, 84–83). Mike spent much of the return trip ripping the wire mesh off the back of his trailer. However, Mike V usually cooperated when it was time to load into his trailer. In this regard, he was easier to work with than his predecessors.

New Wheels for Mike

By September 1997, the trailer used to transport LSU's mascots for forty years—since Mike II—had worn out. Central Hitch and Equipment in Baton Rouge graciously offered to donate a new trailer to LSU. The new trailer was painted LSU gold with purple lettering according to a design submitted by Al Carter. The bars making up the cage portion of the trailer were painted white to reflect heat on sunny days. Many fans wanted the bars painted black for increased visibility of the tiger, but that could result in Mike overheating

LSU cheerleaders atop Mike VI's trailer for his first game in Tiger Stadium. (Wendy Day)

on hot days. Thanks to a lot of hard work by the crew of Central Hitch and Equipment, the new trailer was ready in time for the home opener against the University of Texas–El Paso on September 6, 1997. One need only imagine the dangers of being stranded with a broken-down trailer and a 400-pound tiger to appreciate the importance of a sturdy trailer for the safety of Mike and his handlers.

On January 1, 2002, Mike V once again ventured to the Superdome in New Orleans, this time to support the football team in its game against the University of Illinois in the Nokia Sugar Bowl. Unusually cold temperatures had been predicted for the week. To protect Mike from the cold, his caretakers bedded his trailer deeply with straw and secured it with a custom-made cover. The back was closed up with sheets of plywood. I even took a test ride in the trailer to make sure there were no chilling drafts coming in.

Shortly after we left Baton Rouge, freezing rain began to fall. Mike stayed for the first half of what was turning into a very long game. Not wanting to risk being involved in an accident if the roads froze, we elected to leave ahead of the crowd, which proved to be a wise decision. It snowed nearly all the way back to LSU. Mike reached the campus just as the game was ending in a 48–34 victory for the Tigers. He had enjoyed a long and eventful day, but rather than going directly into his house to rest, he first investigated the soft, cold white powder that coated his entire enclosure. This was probably Mike's first experience with snow. He cautiously smelled it, tasted it, pawed at it, and slipped on it. After several minutes of exploring this new substance, he lost interest and retired to his warm house for a good night's sleep.

Starting in 2015, Mike was no longer taken to sporting events and other venues. Mike's travels have been restricted to medical checkups at the LSU School of Veterinary Medicine. While everyone enjoyed taking Mike to various events, it is better for Mike to remain in his habitat.

Mike VII enjoys a beautiful day. (Katie Miley)

CHAPTER 4

Q&A ABOUT MIKE

It is estimated that Mike receives about 100,000 visitors per year. Here are some of the most frequent questions that people ask about him.

What does Mike eat? Mike is fed a commercially prepared diet made especially for large cats. It looks and feels like hamburger but is made of beef, fish, vegetables, vitamins, and minerals. Mike also receives extra vitamins. Mike is fed fifteen to thirty pounds of food per day. As he ages, he is fed less to keep him at a healthy weight.

Is Mike ever given the chance to hunt and kill live prey? No. It is not necessary to use live animals when feeding Mike. We would not want the prey animals to suffer in the process of being hunted and killed. Also, it is possible that Mike could be injured while killing an animal.

Do Mike's caretakers brush his teeth? Mike's caretakers do not brush his teeth, but they give him an oxtail every week, which helps keep his teeth clean. The oxtail is a bone together with some cartilage and muscle tissue. Chewing on the oxtail helps Mike maintain jaw strength and provides environmental enrichment, as well as keeping his teeth clean.

Does Mike purr? Does he meow? No. Tigers do not make the same sounds as domestic cats. Tigers make four sounds. A chuff (or prustin) is a happy sound and is used in greeting. The moan reveals a minor concern or anxiety. A tiger growls as a warning to other tigers (or to his

opposite Mike VII relaxes on the temperature-controlled rock in his habitat. (Bobby Gilboy)

below Mike VII eats an oxtail in his night house while the landscaping crew spruces up his yard. (Katie Miley)

Mike VII keeps cool.
(Katie Miley)

veterinarian). Finally, a roar is used as an aggressive warning or to locate other tigers in the wild.

Do Mike's caretakers give him baths? No. Mike grooms himself regularly, like other cats. Also, Mike spends a lot of time in his swimming pool. In addition to cooling him off on hot days, this helps his body stay clean.

Is Mike neutered or declawed? No. There is no valid medical reason for either of these procedures. Having all of his body parts allows Mike to develop normally and to express normal, species-specific behaviors other than breeding.

Does Mike have a scratching post like house cats?

Mike uses the oak stump in his yard to keep his claws sharp and at the right length. He pulls his claws along the stump much like a house cat with a scratching post. One time, when the stump fell over, Mike V would not use it anymore. As a result, one of his claws grew too long and had to be trimmed. Since then, Mike's caretakers have always made sure that his oak stump is at the right angle for him to use.

Does Mike get too hot in the summer? Does his night house have air conditioning? It gets very hot in the summer in the areas where tigers are found in the wild. So Mike does fine in the Louisiana summers. Mike stays cool in the heat by lying in the shade and by getting into his pool. In addition, the flat rock in his enclosure contains coils that are cooled in the summer. His night house is also air conditioned.

Does Mike get too cold in the winter? Does he have a heater in his night house? Some wild tigers live in areas where it gets very cold in the winter. While it doesn't get quite that cold in Louisiana, Mike's coat gets thicker as the weather cools in Baton Rouge in the fall. This keeps him warm in the winter. The flat rock in his enclosure

left Mike VII pulls his claws. (Katie Miley)

right Mike VI out in his yard on a rare snow day in Baton Rouge. (Christine Russell)

Mike VI naps.
(Christine Russell)

also has coils that are heated in the winter. Mike also has a heater in his night house.

Mike always seems to be sleeping when I visit his habitat. Why is that? Tigers are nocturnal animals, so they generally sleep much of the day and are more active in the night. The best time to see Mike active in his habitat is early in the morning and at dusk, before the sun goes down. It is natural for cats, including tigers, to sleep up to about sixteen hours per day.

Why does Mike spend so much time walking around his yard? Wild tigers are known to travel up to forty miles per day patrolling their home range. This is especially true for male tigers. They want to make sure that there are no other male tigers in their area, and they want to know if there are female tigers around. So, even though he is in a much smaller area than a wild tiger, this behavior is completely normal for Mike.

Why doesn't Mike sleep outside at night? Animals need a period of total darkness each day for good health. Because of the outdoor lights on campus, Mike needs to be kept in his house at night to benefit from complete darkness. Keeping him inside also protects him from any mischievous guests.

Is Mike bothered by crowds? Mike is a calm and confident tiger who seems to enjoy being around people. So he is not bothered by crowds. He does seem to dislike loud music played near his habitat and other loud sounds.

Does Mike get bored? Mike's habitat was designed to include a variety of views and activities. In addition, Mike seems to enjoy watching people as they pass in front of the enclosure. Also, Mike's caretakers give him toys and place interesting scents around his habitat. Finally, Mike interacts daily with his caretakers and responds to people he has come to recognize. All these factors work together to enrich his environment and minimize boredom. It is not like being in the wild, but then Mike also does not have to deal with all the challenges of the wild, such as diseases, shortage of food,

Mike VI checks out his fans. (Ginger Guttner)

having to kill his prey, and having to compete with other tigers for food and space. Mike's general health and behavior suggest that his habitat is sufficiently interesting and that he is not bored.

Does Mike recognize his caretakers? Mike certainly does recognize his caretakers. He identifies them by sight, voice, and by the way each of them imitates his "chuff" sound. Both Mike I and Mike V knew the sound of their caretaker's car door closing.

Mike VI knew one of his caretakers by the scent of her perfume. Like many other mammals, tigers utilize a special organ near the nasal cavity to detect scents on the molecular level. This is known as the flehmen response. Mike appears to be grimacing when he is actually transferring scent to this delicate organ.

Does Mike miss being around other tigers? Tigers are naturally solitary animals. It is not normal for adult tigers to live together, though if they know each other from their youth they can adapt to such a living arrangement. Tigers are solitary animals largely because of the way they capture their prey. Unlike lions, who hunt cooperatively, tigers, like most other species of cats, are lone ambush hunters. They do not want other tigers

Mike VII bares his teeth while exhibiting the flehmen response. (*opposite and left:* Katie Miley; *right:* Christian Folk)

around who might scare off their prey. In the wild, tigers get together only to breed. The ranges of a few female tigers will overlap the range of a single male, while the ranges of the males do not overlap. It can be very dangerous to introduce one adult tiger to another. There would certainly be a terrible fight, and if one were a female, she would probably be killed.

Have any of the Mikes been female? No. By tradition, all the tiger mascots at LSU have been male.

Why doesn't LSU breed its tiger so that one is always available? There are many reasons for LSU not to establish a tiger breeding program. First, a breeding program would require two or more tigers to live in close proximity, and adult tigers are ordinarily solitary animals. Also, having more than one tiger increases risk and liability for the university. In addition, providing the proper care for tigers is expensive. Supporting multiple tigers is not the best use of limited state or institutional resources. Another consideration is the current surplus of tigers available in captivity. It makes more sense to provide a home for a tiger that really needs one, such as the cubs that became Mikes VI and VII, rather than contribute to the overpopulation of captive tigers. Lastly, there is no way to know pre-

Caretaker Nick Fleissner with Mike VI. (Christine Russell)

cisely when another Mike will be needed, so a replacement could be "waiting in the wings" for several years.

Have any of the Mikes been related to one another? As far as we know, none of the tiger mascots have been related. Other than the second Mike II and Mike III likely coming from the same zoo, they have all come from different sources without any indication that they were related.

Does Mike know his name? There is no way to determine whether Mike knows his name. He certainly responds to the sound of his caretakers' voices, but he does so regardless of whether they call his name or are just talking to him or to others in the vicinity.

Why doesn't Mike ride around Tiger Stadium in his trailer before football games anymore? At one point Mike stopped voluntarily entering his trailer before the football games, after the decision was made to launch fireworks next to his trailer while in the stadium. As his veterinarian, I felt it was important not to force Mike to do things he was uncomfortable with. His safety and well-being are the primary responsibility of his veterinarian and other caretakers. Because of the location of Mike's habitat near the football stadium, all fans can easily visit Mike on game day while he remains in his familiar enclosure.

Why doesn't Mike go to Mardi Gras parades and other social events as he used to do? Taking Mike out on the highway in his trailer entails a certain amount of risk. In fact, over the history of the tiger mascot program, the tiger trailers have been involved in several accidents. There is even more traffic on the roads now than there was back then. For that reason, Mike is now transported off campus only for a medical reason, such as when Mike VI traveled to the Mary Bird Perkins Cancer Center in Baton Rouge for radiation therapy.

Mike VII has plenty of room to run in his enclosure. (Katie Miley)

CHAPTER 5

THE EARLIER MASCOTS

Before the First Tiger Mascot

Some believe that the second president of LSU, David F. Boyd, suggested adopting the tiger as the symbol of the university and naming the football team the Tigers. Boyd had been the commanding officer of the 9th Louisiana Infantry Regiment during the Civil War, a sister unit to an infantry company that called themselves the "Tiger Rifles." Later, the name Louisiana Tigers, and then Lee's Tigers, was applied to all the Louisiana infantry in the Army of Northern Virginia.

Others say that Charles E. Coates, an LSU professor who founded the LSU football program in 1893, should receive credit for naming the football team the Tigers. Dr. Coates was a chemistry professor from Baltimore who served as the head football coach during the team's maiden year. The "season" of 1893 consisted of one game against Tulane University, which Tulane won 34–0. Though off to a slow start, the football program gathered steam rapidly, posting a 2–1–0 overall record in 1894. The following year, under Coach Albert P. Simmons, the team had its first undefeated season, going 3–0–0. Flush with pride and excitement generated by the young football team's success, the student body requested a nickname for the team. The Coates family says that it was Dr. Coates who selected the name "Tigers."

The tiger theme caught on instantly, though for almost forty years other mascot names were batted around, including "Gators," "Roosters," and "Babies" (!). But with each passing year, the tiger theme became more

opposite Mike VI. (Tim Morgan)

Papier-mâché tiger created for the 1920 football game against Arkansas, before the days of the live tiger mascot. (Clarence A. Ives Jr.)

accepted. The name "Tigers" was modified in 1955 to "Fighting Tigers." However, most fans still refer to the team as the Tigers.

From 1896 to 1924, LSU fans carried papier-mâché tigers to the football games. Most of these were destroyed by opposing fans nearly as fast as they were displayed. In 1924, an alumnus donated a small black South American cat, probably related to the bobcat, to LSU. "Jerry" was LSU's first live mascot. Shortly after his arrival, his name was changed to "Little-eat-'em-up" after an LSU football player bit off a portion of an opposing player's ear! Unfortunately, the football team lost all of its conference games that year, so "Little-eat-'em-up" was sent away, and the fans went back to using papier-mâché tigers.

Mike I (1936–1956)

In 1934, a trainer in the LSU Athletic Department and an intramural swimming coach spearheaded the effort to bring a real tiger to LSU. Chellis "Mike"

Chambers and Hickey Higginbotham collected 25 cents from each student, raising a total of $750 over two years. With this sum they purchased Sheik, a 200-pound, one-year-old tiger housed at the Little Rock Zoo. On October 21, 1936, in preparation for the tiger's arrival, LSU President James M. Smith declared a holiday and closed the university. Students paraded through Baton Rouge and onlookers lined Highland Road, awaiting the entourage arriving from the train station.

That evening, the students lit a bonfire and had a grand party. The LSU band played all the versions they knew of "Tiger Rag." By overwhelming vote of the student body, Sheik's name was changed to "Mike" in honor of Chambers, the man most responsible for bringing the tiger to LSU. Just three days after his arrival, Mike I traveled to Shreveport to help LSU defeat Arkansas 19–7. There was a rousing half-time show in Mike's honor. The increased excitement for LSU football was noticeable to all.

At least twice, students celebrated Mike I's birthday by throwing large parties. In 1941 a musical revue, "Hold That Tiger," was scheduled to celebrate Mike's birthday and to coincide with a rally for the LSU–Texas A&M game. Of course, Mike I was the guest of honor. Then, in 1953, the student

LSU athletic trainer Mike Chambers with his namesake, Mike I, at the City Park Zoo in Baton Rouge. (Pop Strange)

Mike I paces in his cage.
(© 1956 *New Orleans Times-Picayune*)

body declared the weekend of October 10 "Mike's Birthday Party." The theme of the party was "Knock out Kentucky for the love of Mike." Mike I enjoyed a large birthday cake, with a raw roast for the bottom layer. The celebration was campus-wide.

As Mike I passed middle age, many were concerned that time was growing short for him to sire a cub. After all, he would eventually need an heir. Two attempts were made to send him on a quest for love at the Audubon Zoo. However, in each case the LSU student body blocked the efforts. They did not relish the idea of their beloved mascot traveling to New Orleans, the home of their archrival, the Tulane Green Wave. Finally, in July 1945, Mike I traveled to the R. M. Taylor Zoo in Jackson, Mississippi, to meet a lovely 3½-year-old female tiger named Desdemona. Keepers introduced the two tigers to one another through fencing. Mike I stayed at the Jackson Zoo for a few months, but the two tigers never really hit it off. The risk of injury to one or both of the tigers was too great to allow contact, so Mike I returned to LSU still a bachelor.

On June 29, 1956, Mike I died of complications associated with kidney disease. In backyards all over Louisiana, children held mock funerals in his honor. Others hung his photograph, draped in black crepe paper. Many people who had been students on campus during his tenure or who had grown up visiting him mourned his passing. He was twenty years and eight months old at his death and had reigned at LSU for nearly twenty years. Without a doubt, he established "Mike the Tiger" as the heart and soul of LSU. Following Mike's death, a fund was established to perpetuate his memory by mounting his pelt in a lifelike manner for display in the LSU Museum of Natural History, where he proudly remains to this day, patriarch of a long and great LSU tradition.

MIKE I'S STUDENT CARETAKERS

- 1936–37: L. C. Aycock
- 1937–38: Preston Vallas
- 1938–39: Elliot "Eddy" Contine
- 1939–42: Charles Fourrier
- 1942–43: Osce R. Jones
- 1943–44: Charles Upp
- 1944–45: George Thompson
- 1945–46: E. Loy Upp
- 1946–47: H. Staton Barlow
- 1947–49: Brittain Briggs
- 1949–51: David Melilli
- 1951–53: Jere D. Melilli
- 1953–55: Bob Dodwell
- 1955–56: Gene Smith

Mike II (1956–1958)

On July 1, 1956, the Louisiana legislature introduced a resolution conveying sympathy to the LSU community for the loss of Mike I and endorsed the purchase of another tiger. The "Mike the Tiger Fund" was launched on campus with the goal of finding a new tiger by the start of the next football season. Fortunately, the money was raised quickly. Meanwhile, the LSU Board of Supervisors determined that caretaker salaries and maintenance costs for the new tiger would come equally from student fees and the LSU Athletic Department. This was an important step to assure proper care of Mike II, as well as his successors.

LSU's second tiger was born on February 28, 1956, at the Audubon Zoo in New Orleans. He arrived secretly on the LSU campus on September 28 of that year and was introduced as Mike II the next day. The university declared an academic holiday, and the celebrations began. A parade led by the LSU band wound through downtown Baton Rouge. Later, there was a general campus open house, a drill routine, a salute by the Pershing Rifles, and a welcoming ceremony honoring the new, regal addition to the LSU family.

To this day, no one knows for sure what happened with Mike II. Legend has it that, less than a month after his arrival at LSU, he died of pneumonia. According to Jack Gilmore, who was the LSU Athletic Department business manager at the time, a campus police officer making his rounds found the tiger dead in the early morning hours. The timing couldn't have been worse—the football team was in the throes of a 6-game losing streak, and morale on campus was already low. So, reportedly, Mike II was secretly buried under a tree on the Mississippi levee by Gilmore, the campus police chief, and the newly appointed athletic director, Jim Corbett. To explain Mike's absence from public view, the LSU administration issued a statement that Mike was having trouble adjusting to his enclosure and was therefore being kept inside "until he becomes more accustomed to the excitement of being a mascot."

Corbett's son, Jim Corbett Jr., was eleven years old at the time. He has said that his father never told him what happened next, but he recalls phone calls being made in a frantic effort to find another tiger. Gilmore located a cub of the right age at the Woodland Park Zoo in Seattle and arranged for the cub to be delivered to the Audubon Zoo.

In addition to Gilmore's testimony, evidence supports the legend that Mike II died and was secretly replaced by another young tiger. First, records

A young Mike II.
(Philip Guarisco, © 1956 *New Orleans Times-Picayune*)

from the Woodland Park Zoo in Seattle list a litter of three cubs born on almost the same date as LSU's Mike II. There is no notation on what became of those cubs, but Audubon Zoo records from 1956 refer to a tiger cub purchased from the Seattle Zoo and transferred to Baton Rouge. Second, photographs taken of Mike II before and after his period of acclimation are clearly of two different tigers. Careful examination reveals that the facial markings of the two tigers differ. Third, in November, when Mike was once again allowed to leave his tiger house, he seemed to have grown at a tremendous rate. Records from the time cite Mike's weight at 65 pounds in September 1956 and 130 pounds just two months later.

Rumors of Mike's death and replacement swirled around the university,

The second Mike II.
(Erby M. Aucoin Jr.)

but they were firmly denied. Various persons in authority stated that Mike II had finally adjusted to life as a mascot and was in good health.

Imposter or not, the Mike II who returned to the public eye in November 1956 reigned at LSU for only one season. In September 1957, he broke his left rear leg in multiple locations, perhaps from jumping to the concrete floor of his cage from a pedestal above. Mike II was taken to the Audubon Zoo the following spring for treatment but died from an infection on May 15, 1958.

At least one very positive development came during Mike II's short reign. In September 1957, a new, custom-built "tiger chariot" was purchased. This "tiger taxi" replaced the much smaller and less stable cage that had been used to transport the tiger mascot. The new trailer would serve as Mike's traveling cage for the next forty years. When it was replaced in 1997, the same design was used with only a few modifications.

MIKE II'S STUDENT CARETAKERS

- 1956–57: Gene Smith
- 1957–58: Sidney S. Miller
- 1958: Bill Levenson

Mike III (1958–1976)

Just in time for the National Championship football season of 1958, Jim Corbett and Jack Gilmore located and purchased Mike III, coincidentally, from the Seattle Zoo. The newest addition to the LSU tiger family was born on November 26, 1957, and arrived at LSU in late August 1958. The LSU student body raised $1,500 for his purchase and transportation. Mike III was flown to Chicago and then on to New Orleans. His first student caretaker, Bill Levenson, reported that the new mascot adjusted easily to life at LSU. Mike III was introduced to the public at the first home game of the 1958 season, which LSU won 20–6 against Hardin-Simmons University.

Mike III lived a long and mostly healthy life. However, shortly after arriving at LSU, he broke both of his rear legs, probably while jumping onto the hard cage floor. Fortunately, he was moving around and active again after about six weeks, and his limbs healed quickly under the care of Dr. Joe Dixon of the LSU Department of Veterinary Science. Douglas Davidson, Mike's caretaker from 1959 to 1961, reports that, before one LSU-Tulane football game, some Tulane students threw green paint on Mike III. However, no harm came to the young tiger. After that incident, the caretakers would lock Mike up for several days before the LSU-Tulane football game to protect him from pranksters.

Mike III challenges the photographer. (Joel Samuels)

Mike's old enclosure, in 1975. (*Baton Rouge Advocate*)

In 1976, with Dr. W. Sheldon Bivin now serving as veterinarian for the tiger mascot, Mike III showed signs of age. He had developed arthritis, which made simple things like walking and even eating difficult. Some suggested that Mike be humanely "put to sleep" but others objected, and so Mike lived on for a short while.

Around this time, efforts to provide Mike with a larger, more naturalistic enclosure gained steam. Dr. Bivin suspected that Mike III would likely not live long enough to see a new enclosure, but he supported the effort, knowing that future mascots would benefit from an improved, more natural habitat. Mike III reigned 18 seasons at LSU. During that time, LSU won a National Championship (1958) and 3 SEC championships (1958, 1961, 1970) and played 13 bowl games, winning 8 of them. Mike III died of pneumonia on August 12, 1976, after the only losing season of his reign.

MIKE III'S STUDENT CARETAKERS

- 1958–59: Bill Levenson
- 1959–61: Douglas Davidson
- 1961–63: Alvin Meyer Jr.
- 1963–65: Paul Marks Jr.
- 1965–68: Joel Samuels
- 1968–72: Robert Lank Jr.
- 1972: Philip B. Aman
- 1972–76: Randy Kilgore
- 1976: Sonny Corley

Mike IV (1976–1990)

Mike IV was born at Busch Gardens in Tampa, Florida, on May 15, 1974, and was donated to LSU by August A. Busch III. "Jerry" was hand raised when his mother abandoned him shortly after birth. Despite all his early contact with people, Mike IV turned out to be the most aggressive of the LSU tigers to date. The two-year-old, 450-pound tiger arrived at LSU on August 29, 1976.

Five years after coming to LSU, Mike IV spent the summer of 1981 at the Little Rock Zoo while his enclosure was expanded and updated. Mike returned from Arkansas to find a swimming pool and real grass in his yard. After a period of adjustment, Mike came to enjoy his new digs, especially his pool.

Mike IV's expanded enclosure. (Jim Zeitz, University Relations)

TIGER ON THE LOOSE

November 28, 1981, will be forever remembered as the day Mike the Tiger got out of his enclosure. At about 1:00 a.m., Dr. Bivin was woken out of a sound sleep by the LSU police. "Mike's out," they said. "Mike who?" he asked groggily. As it happened, Dr. Bivin was not only sleepy but suffering from flu symptoms. The police responded, "Mike the Tiger . . . he's out in the middle of North Stadium Drive."

How could Mike possibly have made "the great escape"? Apparently some pranksters had cut the chain to the outer door of Mike's cage and also the lock on the inner cage door, releasing the dangerous cat. Once Dr. Bivin was convinced that this was not a joke, he retrieved his rifle and rushed to campus. There he picked up tranquilizers and a dart gun. By the time he arrived on the scene, Mike was wandering toward the Pete Maravich Assembly Center. He attacked a small tree before entering the Bernie Moore Track Stadium. Police positioned their cars in such a way to block Mike within the stadium and to shine their lights toward him. Dr. Bivin then bravely entered the stadium with his tranquilizer pistol in one hand and his rifle in the other. The rifle was for his own and others' protection.

Dr. Bivin managed to get downwind of Mike. The problem was that he had to be within about forty feet of the tiger to get a clean shot with the tranquilizer pistol. That's closer to Mike than a pitcher is to a catcher! Bivin got as close as he could and fired one dart into Mike, who bolted and ran a short way, then spun around to see where the dart came from. He let out a growl and then a roar. He was mad! Within a few minutes the tranquilizer took effect, and his back end became uncoordinated. Then Dr. Bivin was able to get close enough to get a second dart into him. Mike bolted down to the north end of the stadium and stood there for a while before slowly going down, growling. That's when Dr. Bivin went right up to him and injected him with a third dose. When Mike IV was completely tranquilized, Dr. Bivin backed his truck up and, with help from the police, lifted him into the truck for the short ride back to his enclosure. Mike awoke the next morning with no ill effects.

Mike IV reigned for fourteen years at LSU and was retired to the Baton Rouge Zoo in April 1990. Mike had developed a neurologic condition resulting in mild lameness. In addition, he was getting along in years and was beginning to slow down. The zoo graciously offered to take in the aging mascot and house him in a private enclosure out of public view. There he lived

opposite An angry Mike IV. (Mark Rayner)

Mike IV on the raised platform in his enclosure, ca. 1985. (Harry M. Cowgill)

until his condition worsened and he became severely disabled. Mike IV was humanely euthanized on March 3, 1995. His life span of 20 years, 9 months, and 18 days is one of the longest on record, and the longest yet for any of the LSU tiger mascots. The normal life span of a tiger is about 8–12 years in the wild and around 14–18 years in captivity.

After his death, it was discovered that Mike IV had pneumonia and also cancer of his thyroid gland. The cancer had spread into his chest and liver. That finding confirmed that putting Mike IV to sleep had been the only humane option. Mike IV was cremated, and his cremains were eventually put on display in the LSU Andonie Sports Hall of Fame.

MIKE IV'S STUDENT CARETAKERS

- 1976–78: Sonny Corley
- 1978–80: Sonny Corley and Donald J. Harris Jr.
- 1980–82: W. Byron Garrity Jr.
- 1983–84: John R. Allender
- 1985–86: John D. DeVun
- 1987–88: Jeff Larisey
- 1989–90: Jeffrey Perret

Mike V (1990–2007)

Born October 19, 1989, Mike V was donated to LSU by Dr. Thomas and Caroline Atchison of the Animal House Zoological Park in Moulton, Alabama. He was introduced to LSU fans at the LSU-Alabama basketball game in February 1990. One of three cubs in his litter, Mike was originally named "Stevie." Apparently, being a mascot ran in the family; Tom II, a brother of Mike's from a later litter, served as the proud mascot of the University of Memphis until his death in 2008.

Mike V had a calmer, friendlier disposition than most tigers and so was relatively easy to care for. This was in large part because he came to LSU at just four months of age and was hand raised by loving caretakers. When he was young, he liked to stand up and grasp people by the back of the neck with

A young Mike V relaxes in his small yard. (Harry M. Cowgill)

Four-and-a-half-month-old Mike V plays with Fran Lupp and a toy tiger. (*Baton Rouge Advocate*)

his jaws or dunk them under in his swimming pool. When his weight topped 250 pounds, people were no longer allowed in his enclosure with him. At that weight, Mike might accidentally hurt someone. And because he was a tiger and not a domestic animal, a day would probably come when his wild instincts would emerge, and he might intentionally attack someone.

I became the third veterinarian for LSU's tiger mascot when Mike V was six years old. Mike's reign was punctuated with a highly successful collegiate sports program: 29 national championships in six sports. Most notably, the women's track-and-field teams won an amazing 19 national championships between indoor and outdoor venues. During the same period, the men's track-and-field teams clinched four national championships, and the baseball team won five national championships. One of the most exciting victories during the reign of Mike V was LSU's 31–14 defeat of Oklahoma on January 4, 2004, to win the BCS National Championship for the 2003 football season. It was LSU's second national championship in football in the school's history.

In the spring of 2007, at seventeen years of age, Mike was noticeably contending with age-related health issues. For example, he appeared to have developed sensitivity in his lower incisors, which had worn considerably over the course of his long life. This resulted in a decreased appetite and contributed to a loss of nearly seventy pounds of body weight. On May 15, his caretakers observed that Mike was having trouble breathing. He was fasted overnight in preparation for anesthesia the following morning. After receiving the anesthetic, he was taken to the LSU School of Veterinary Medicine, where we saw that he had a considerable amount of fluid on his lungs. We could not determine the source of the fluid, but we removed it. Unfortunately, Mike's aged kidneys failed, unable to handle the reduced blood flow that often occurs under anesthesia. Mike underwent renal dialysis (a historic first in tiger medicine) and we took every other step possible, but his kidneys simply would not resume proper function. On May 18, Mike V died. His body was cremated later that day and donated to the LSU Andonie Sports Hall of Fame.

It was a particularly hard day for the LSU community to lose Mike V because it was also graduation day. Many new graduates, their families, and their friends visited Mike's enclosure expecting to see him lounging in the shade but instead found a floral wreath signifying his passing. The Athletic Department set up a website so that Mike's admirers could send in their remembrances. Nearly 750 messages were received. Governor Kathleen Blanco expressed the feelings of many when she said, "Mike V served LSU as more than a mascot. He was an epic symbol that represented the unique spirit of our state's flagship university."

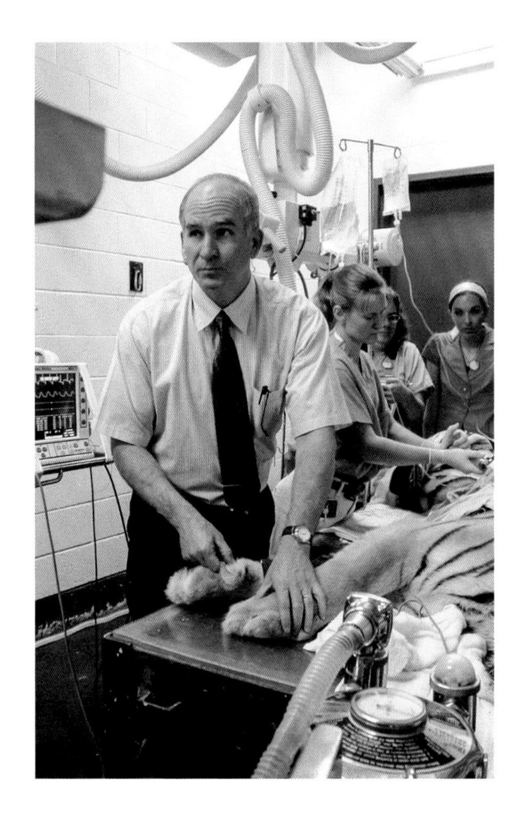

Mike V shortly before his kidneys failed. (Ginger Guttner)

Caretakers Wendy Day and Wesley Lee with Mike V. (Sandy Oubre)

MIKE V'S STUDENT CARETAKERS

- 1990: Jeffrey Perret
- 1990–92: Rollie Norris and Shelly Phillips
- 1992–94: Jeffery Artall and Mark Gentry
- 1994–96: Randolph Hayes and Lance Hidalgo
- 1996–98: L. Shane Parker and David B. Webre
- 1998–2000: Stephen Eastman and Leath Harper Jr.
- 2000–02: Kirk Mauer and Matthew Wheelock
- 2002–04: Kevin Bankston and Blake Tunnard
- 2004–06: Nelson Lewis and Timothy "Toby" Wallis
- 2006–07: Wendy Day and Wesley Lee

REMEMBRANCES FROM THE

LSU COMMUNITY

After the death of Mike V, LSU received condolences and anecdotes from more than seven hundred people whose lives had been touched by him. Below are just a few of them.

"How well I remember the day all those years ago at the Assembly Center when they trotted out the little cub that would grow to be Mike the Tiger. . . ." —Scotty, Baton Rouge, LA

"LSU has lost its king!" —Alex, Lake Charles, LA

"To the only Mike I've ever known . . . You implanted the LSU spirit in me." —Lynette, New Orleans, LA

"I was on campus the day that Mike V came to LSU. I remember him playing and tumbling around like it was yesterday. To me, he will always be the little cub that poked his nose in everyone's face and warmly greeted every admirer. . . ." —Scott, Houma, LA

"You were the fiercest and most magnificent of mascots." —Janet, Miramar Beach, FL

"Mike V was our tiger!!"—Annie, Baton Rouge, LA

"Coming to my first LSU game and seeing Mike V in his habitat was the reason I came to LSU." —Michael, Franklin, LA

"My son learned to spell T-I-G-E-R-S before he could spell his own name thanks to Mike!" —Rodney, Little Rock, AR

"I was married in front of Mike's habitat, so he was the first one there for the ceremony!" —Jami-Lynn, Baton Rouge, LA

Mike V at his first LSU basketball game in 1990. (Patrick Dennis, *Baton Rouge Advocate*)

"I was in attendance at the basketball game where Mike V was introduced. It was electric!" —Clyne, Walker, LA

"Your roar sent chills through many . . . your death brought tears to all." —Renee, Baton Rouge, LA

"There have been many tigers at LSU, but you were my Mike." —Ashley, Texarkana, TX

"Thank you, Mike V, for making my experience at LSU the best thing that ever happened to me." —Joshua, Raceland, LA

"Mike V was the heart and soul of LSU. He defined LSU and will be sorely missed." —Jamie, Lafayette, LA

"Mike V, you were the most magnificent and magical animal I have ever seen." —Rita, Shreveport, LA

Mike VI relaxes in front of the waterfall in his habitat. (Christine Russell)

Mike VI (2007–2016)

The search for Mike V's replacement began the day he died. We hoped to obtain a new tiger from a licensed rescue facility rather than a zoo or animal breeder. I took advantage of the Internet, which was not available in previous searches, to quickly pull together information about available tigers without having to visit all of them. A veterinary colleague with the USDA told me about Great Cats of Indiana, a privately run rescue facility near Idaville, Indiana. Through my initial investigations of the facility, it appeared to be a suitable source. After speaking with the owner, Rob Craig, I believed that this facility had what we were looking for. So, on June 25, 2007, Dr. Gordon Pirie and I flew to Indiana to investigate.

We were introduced to Roscoe, a beautiful two-year-old Bengal-Siberian cross. Compared to Mike V, who was a Bengal-Indochinese cross, Roscoe had a longer muzzle; whiter face and undersides; a larger ruff (which is smaller but somewhat similar to a lion's mane); and more double stripes, which make for a tiger of truly stunning appearance. Born July 23, 2005, Roscoe was the product of a breeding that occurred when his father, Tigger, an enormous Siberian tiger weighing perhaps 700 to 800 pounds, tore down his enclosure fence and bred two female tigers in the adjacent pen. Roscoe's mother, Kiara, was a graceful Bengal-cross tiger with striking markings.

Dr. Pirie and I were impressed with the overall condition of the seventy carnivores residing at the Great Cats of Indiana facility, and it didn't take me long to decide that Roscoe belonged at LSU. The following month, my colleague Dr. Rhett Stout and I flew to Indiana to retrieve Roscoe, courtesy of Acadian Ambulance Service. The trip was a great success. Every step of the relocation process went flawlessly, much to my relief and that of his other caregivers.

On September 1, 2007, Roscoe stepped into his yard for the first time and became Mike VI. He was confident but naturally cautious as he explored his new home. He was very interested in the people gazing at him through protective glass and got as close to them as he could. Until that day, he had

left Preparing the transport crate for Roscoe's flight to LSU. (Rhett Stout)

right Loading Roscoe onto the plane for the flight to LSU. (Rhett Stout)

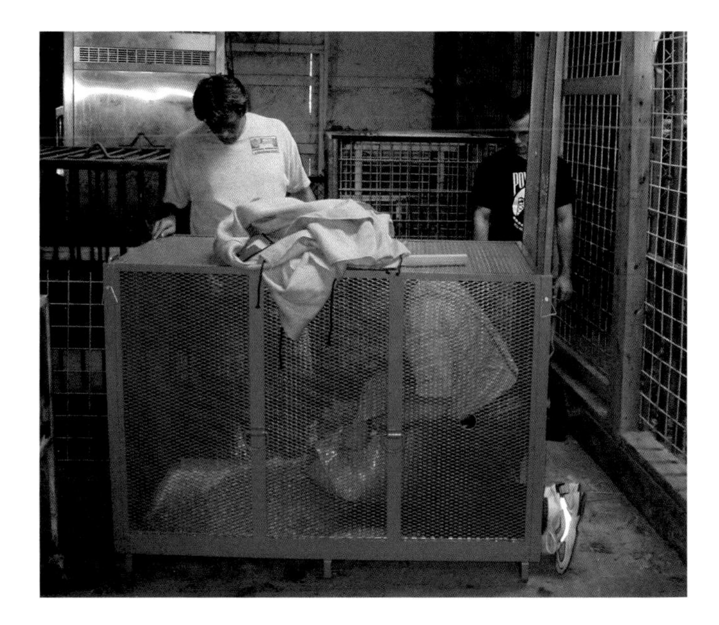

left Mike VI marks his territory on an oak tree in his habitat yard. (Christian Folk)

middle Mike VI has a major battle with his boomer ball. (Robbie Gray)

right Mike VII guards the rings for graduation day, December 2022. (Caitlin Bell)

never seen a plant, nor had he ever seen more water than could fit in a bucket. One can only imagine what was going through his mind as he explored what to him must have seemed like paradise.

The 320-pound Mike VI was an immediate hit with the LSU faithful who showed up to greet their new mascot. In a special ceremony held at his enclosure on September 14, 2007, Chancellor Sean O'Keefe officially welcomed him as Mike VI.

Over the course of his life, Mike VI was an outstanding tiger. Not only was he beautiful, but he also had an affectionate and playful personality. As tigers go, he was easy to care for. Even so, his caretakers never let themselves forget that even a hand-raised tiger is not a domestic animal, and certainly is not a pet. His caretakers had to remain vigilant to prevent him from injuring himself or others.

In 2012, during Mike VI's reign, a new tradition began at LSU. Twice each year, a few days before the fall and winter graduation ceremonies, a large box containing the LSU class rings is placed inside Mike's night house, out of reach, to spend the night with Mike the Tiger. The new LSU graduates seem to appreciate that their ring "spent the night" with Mike.

TUMOR AND RADIATION

On April 30, 2016, just days before turning over the care of Mike VI to his new student caretakers and graduating from the LSU School of Veterinary Medicine, Lindsey Clemones noticed a slight swelling of Mike's face, just under his right eye. She didn't know what might be causing the swelling, but she knew it wasn't normal and couldn't be good. She notified me immediately. After examining Mike, I thought that he might have developed a tooth abscess, not uncommon in tigers. Just a few weeks earlier Mike had broken off a portion of his upper right canine tooth, which supported this presumption. What I found perplexing was that Mike wasn't showing any of the usual signs of a tooth infection; that is, he didn't appear to be in pain, his tooth was not inflamed, and his appetite was normal. I decided to arrange for a computed tomography (CT) scan of his head to determine the extent and cause of the swelling. This occurred on May 12.

The results of the scan were devastating to his caretakers. To everyone's surprise, Mike VI had a large, space-occupying mass in the bones of his sinuses, most likely a tumor. A biopsy specimen was collected from the mass and determined to be a spindle cell sarcoma, a very aggressive form of cancer.

I met with university officials and explained Mike's condition and grave prognosis. I felt strongly that how LSU treated Mike VI in his time of need would say much about LSU. Would we just euthanize him because he was

left Mike VI has a diagnostic CT scan at the School of Veterinary Medicine. (Ginger Guttner)

right Dr. Charles Wood reviews the CT scan while I wait. (Eddy Perez, LSU)

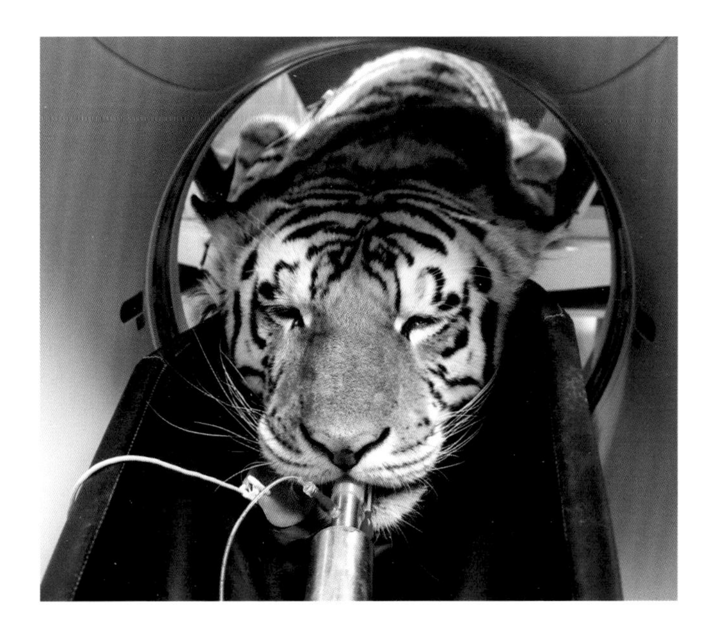

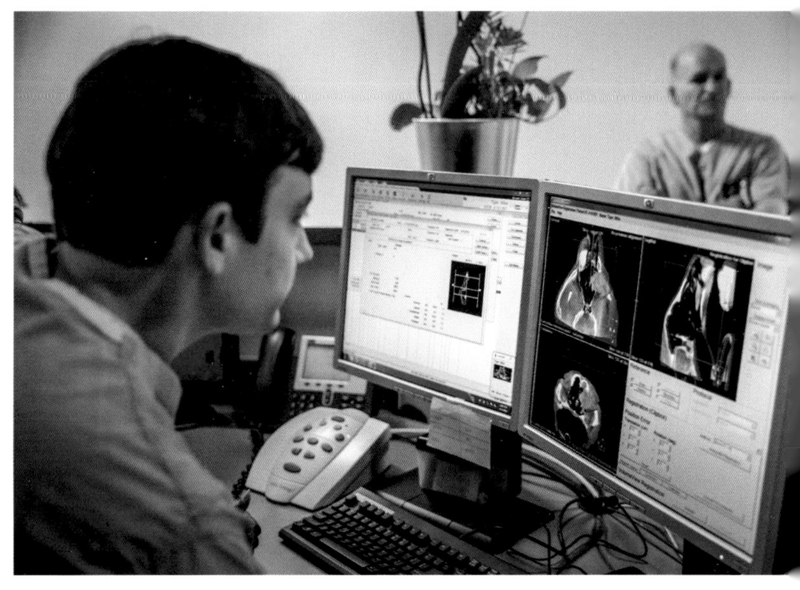

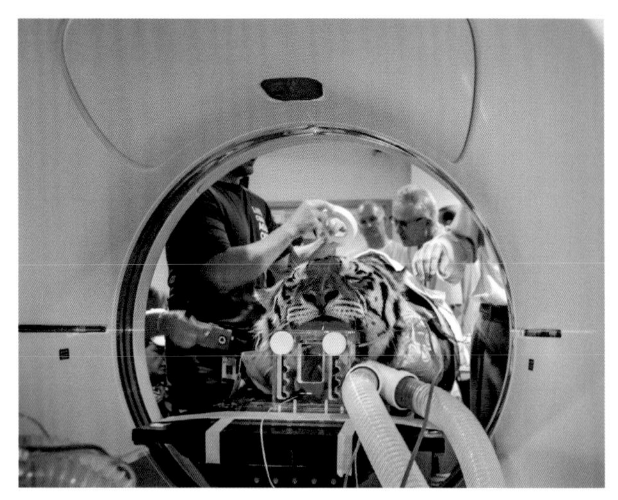

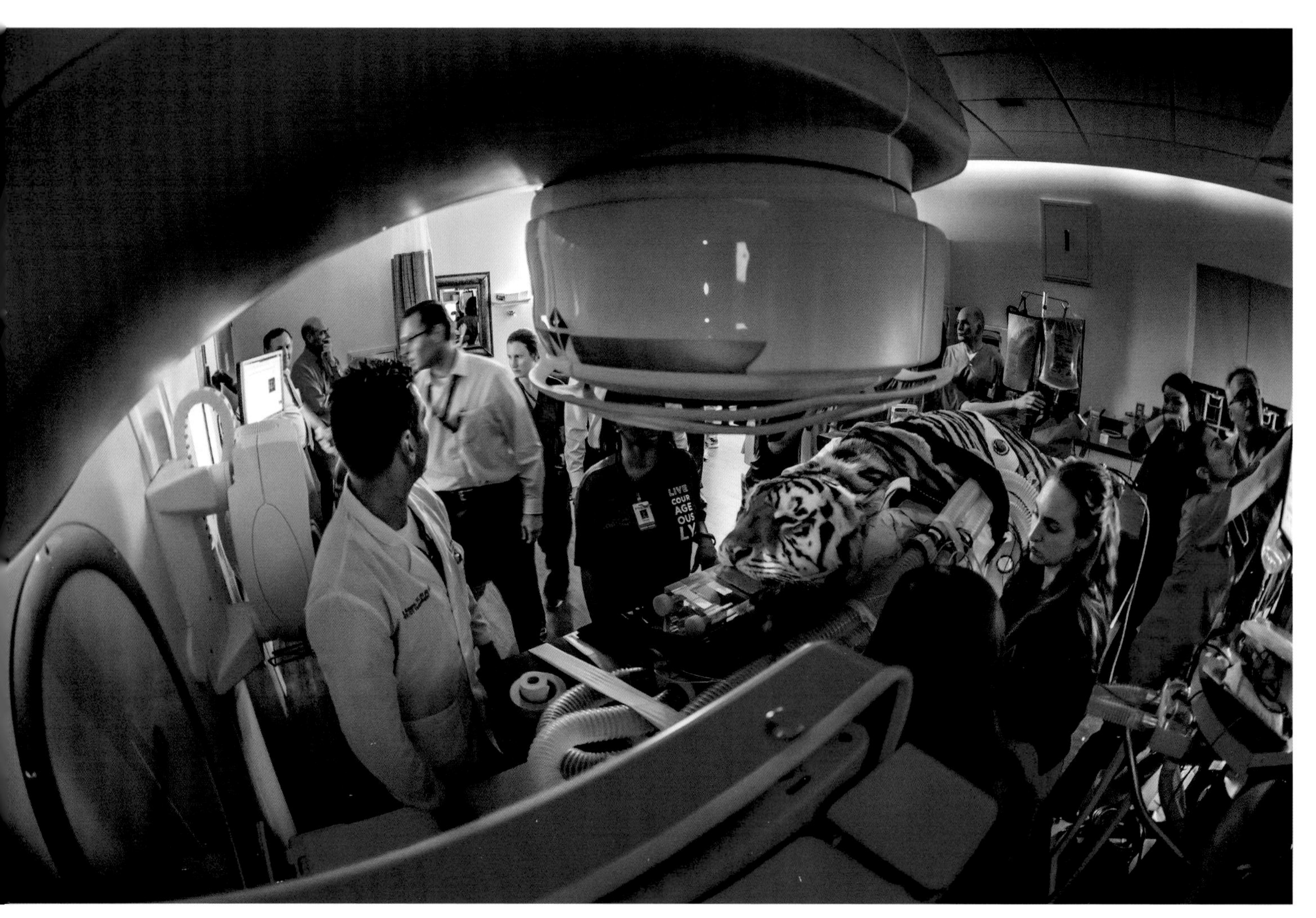
LIVE
COUR
AGE
OUS
LY

Several days prior to his radiation therapy at Mary Bird Perkins Cancer Center, Mike VI underwent a preliminary CT scan. (*opposite page, top left and bottom:* Eddy Perez, LSU; *top right:* Ginger Guttner; *above:* Eddy Perez, LSU)

no longer "perfect"? Or would we provide the best care possible for him out of love and respect for him as a valued member of the LSU community? The LSU administration was adamant that Mike VI be provided the best care available. With that commitment, I assembled a medical team that included professionals from both veterinary and human medicine. Among others, we had a veterinary radiologist, a veterinary anesthesiologist, and a veterinary oncologist from the LSU School of Veterinary Medicine; a human cancer surgeon and a radiation physicist from the Mary Bird Perkins Cancer Center, then affiliated with Our Lady of the Lake Hospital in Baton Rouge; and a veterinary oncologist from the Colorado State University College of Veterinary Medicine.

On May 28, we anesthetized Mike and brought him to the Mary Bird Perkins Cancer Center for a planning CT. This scan was used to establish the

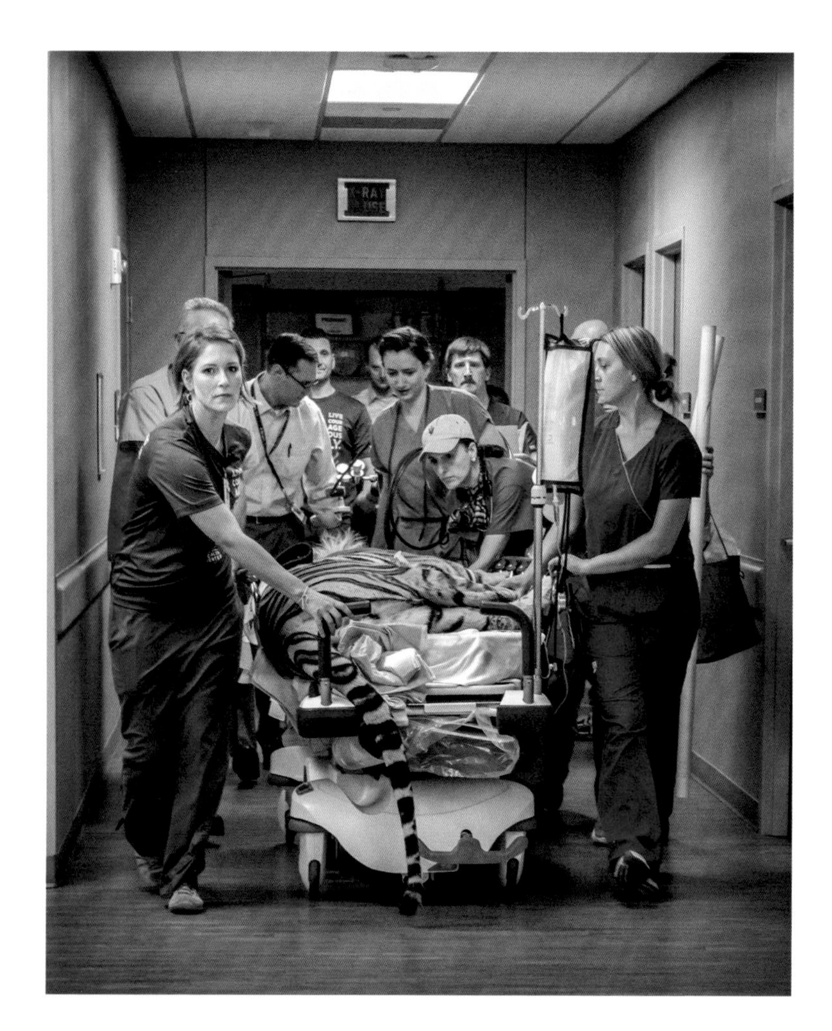

left Mike VI after the preliminary scan. (Ginger Guttner)

right Mike VI shows off the effects of his radiation treatment. (Christian Folk)

parameters for irradiating Mike's tumor using stereotactic radiation therapy, a procedure never before performed on a tiger. Over time, the facial swelling had enlarged and was pressing on Mike's right eye. At the rate the tumor was growing, Mike would soon lose his eye, as well as his life. On June 1, Mike VI was again anesthetized and transported under police escort to the Cancer Center, where he received radiation therapy. The goal of the treatment was not to cure the cancer, since that was not possible, but to give him a few more months of high-quality life. This would also give the LSU community time to say their goodbyes to Mike VI.

The tumor shrank within days of radiation treatment. For the next three months, other than a distinguished "eye patch" from the radiation, Mike VI appeared normal. However, his medical team knew that it was just a matter of time until the tumor resumed growth. By late September, it was apparent that the tumor was growing again. On October 3, when we examined Mike VI at the LSU School of Veterinary Medicine, we learned that Mike's cancer had spread to many other locations, indicating that death was imminent. I assured the LSU community that we would not allow Mike VI to suffer.

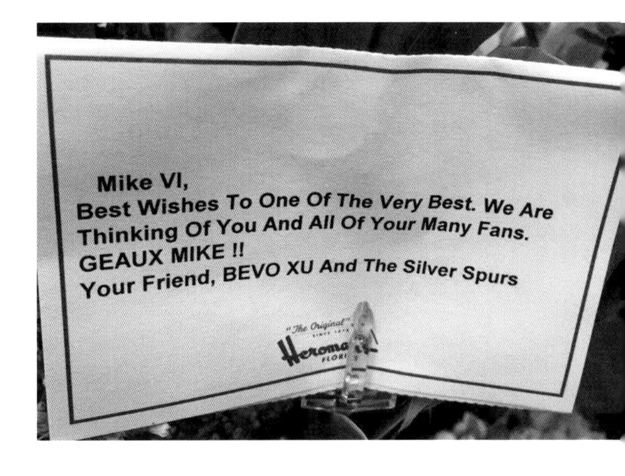

above UT–Austin mascot Bevo, a Texas Longhorn steer, sends his get-well wishes to Mike VI. (Ginger Guttner)

following page Get-well cards and gifts for Mike VI. (Ginger Guttner)

FINAL GOODBYES

Death comes to every living thing. Sometimes, humane euthanasia is the most loving course of action in providing medical care to our animals. The LSU community was informed that as Mike's veterinarian, I would make the decision regarding when the time had come to euthanize Mike VI and that I would perform the procedure. Over the next week, many of Mike's fans came to say their goodbyes. When it was time for the procedure, Mike was sedated and was lying down with his head on his paws. Caitlin Bell, one of his student caretakers, was kneeling directly in front of him. Mike VI raised his head, looked directly at Caitlin, and gave her a final chuff, which is a happy sound of greeting. It was a remarkable moment. Immediately thereafter, he put his head down, lost consciousness, and was euthanized.

The mood was somber throughout the LSU community. The outpouring of support demonstrated how loved Mike VI had been. Many questioned whether it was possible to find such a wonderful tiger to replace him. Fortunately, the new, young tiger we introduced to the LSU community as Mike VII won the hearts of all.

MIKE VI'S STUDENT CARETAKERS

- 2007–08: Wendy Day and Wesley Lee
- 2008–10: Adam Caro and Kelly Folse
- 2010–12: Nicholas Cross and Randee Monceaux
- 2012–14: Daniel Cutler and Macy Trosclair
- 2014–16: Lindsey Clemones and Trevor Davis
- 2016: Caitlin Bell and Christian Folk

get well mike
we you
Country Day School of Baton Rouge
May 25, 2016
Dear Mike,
You are cool! We ♥ you! We know you are sick, but we hope you are better soon. Do you have friends? Be sure to take your medicine, and get some rest.
We ♥ you Mike!
Tiger Tails Purple Group
Audrey
GEAUX TIGERS!
get well soon! We ♥ you!
We ♥ you Mike
We LOVE you Mike the Tiger
Love, Country Day School of Baton Rouge
Get Well Soon mike
Fight like a Tiger, MIKE VI!
Get Well Soon Mike!
To: Mike
WE LOVE YOU, MIKE!
MIKE VI

REMEMBRANCES FROM

MIKE VI'S CARETAKERS

Wendy Day (2007–08): "One of the many blessings in my life is working with animals. Having had the chance to care for Mike V in his last year and welcoming Mike VI to LSU his first year was amazing. Nothing can replace the bond I developed with either Mike, especially the chance to be Mike VI's 'LSU girlfriend.' I still remember his reaction and how I could never sneak up on him. When I drove up, he knew the sound of my car engine, the sound of my door shutting, the jingle of my keys, and the smell of my perfume. Even after I graduated from veterinary school, Mike VI still remembered me when I went by and visited. He will always have a piece of my heart." *Dr. Day works at White Oak Animal Hospital and ER in Baton Rouge, Louisiana.*

Wendy Day gives Mike VI a snack. (Ginger Guttner)

Wesley Lee (2007–08): "Word of Mike VI's cancer hit like a sucker punch to the gut. Mike VI came along as such a bright spot in a dark time after we lost Mike V. His youth and enthusiasm were such contrasts to the wisdom and composure of the veteran old Mike V. There is something special in the experience of having any animal recognize you as an individual and truly be excited to see you . . . and when that animal is Mike the Tiger, 'special' falls woefully short as an adequate description. It was such a privilege to share part of my days with these two amazing tigers. It was wonderful to see Mike VI grow and mature into a beautiful tiger. Sending a heartfelt chuff your way, old buddy." *Dr. Lee works at Red River Equine Hospital in Benton, Louisiana.*

Adam Caro (2008–10): "I will be forever grateful for the time I was able to be Mike's caretaker. The training one receives to become a veterinarian is long and difficult. Classes begin early in the morning and go until late in the afternoon. Veterinary students are studying constantly and are in labs after hours. Once clinical rotations start, it can become even more hectic. As Mike's caretaker, I was there in the mornings caring for him and maintaining his habitat and quarters and returned in the evenings. Not once did I regret taking on this extra responsibility. Often the best part of my days were the times I spent with Mike VI. Waking him in the morning as he stretched and yawned in his bed of shavings and chuffed at me with his greeting always put a smile on my face and readied me for the day. He loved to play games, his favorite being hide-and-seek as I walked around his habitat looking for anything that needed repairs or tending to. I often spent extra time sitting with him in his night house after his evening meal. It was a quiet time of my day, and I think he enjoyed the company. I will always fondly remember my

Nick Cross checks Mike VI. (Randee Monceaux)

time caring for Mike and working with everyone who takes care of him." *Dr. Caro serves as clinical veterinarian for LSU Health Sciences Center in New Orleans, Louisiana.*

Nicholas Cross (2010–12): "I had been gone for over a year for my internship in Indiana. Upon returning, I paid a visit to my old friend at his enclosure. There was a sizable crowd that day. I stepped out of my car and walked over to where Mike was sitting. As I made my way over and emerged through the crowd, I spotted him. He looked majestic, bathed in sunshine in all his glory. Right then was an amazing moment for me. He picked up his head and gave me a little chuff. After so long, he still remembered me! I couldn't believe it. Mike stood up and began pacing along the fence, following me from one end to the other. Those around us were astonished. Someone asked me, 'How do you get him to do that?' And in reply all I could say was, 'That's my buddy Mikey.'" *Dr. Cross is a small-animal practitioner at Banfield Pet Hospital in Covington, Louisiana.*

Randee Monceaux (2010–12): "It was astonishing how intelligent Mike VI was, at times almost human-like. Although wild tigers are generally solitary animals, Mike VI formed deep and trusting bonds with a few people. I was fortunate to have been one of those people. As one of his caretakers I spent several hours with him each day, providing necessities such as food, veterinary care, and enrichment. Mike and I would often sit in his night house or yard, side-by-side, just quietly enjoying each other's company. This became such a comforting ritual for me that leaving Mike VI after graduation was like moving away from my best friend. Several years later, I returned to Baton Rouge for a football game. Of course, my first stop was the habitat to see my buddy. The front of the habitat was full of tailgaters, making it doubtful in my mind that Mike would even notice me in the crowd. I saw him walking around his habitat with his head held high. He liked to strut in front of his admirers. I 'chuffed' as loud as I could, hoping he would recognize the sound, but he kept walking as it was likely something he heard all day long. So, I yelled out, 'Mikey!' He stopped suddenly in his tracks. He looked over his shoulder, his ears perked up and his eyes searching. I chuffed again, and I saw him focus on me. He immediately leaped around to run up to the fence, chuffing and rubbing his cheeks on the netting, like a house cat would rub its owner's leg in greeting. He stayed there with me for what seemed like

Randee Monceaux and Nick Cross on the field with Mike VI. (Phillip Cancilleri)

an hour. Everyone asked how I was making him do that. I would just reply, 'I'm not making him do anything.' It was refreshing to be back with my friend after so long." *Dr. Monceaux is a small-animal practitioner at the Animal Emergency Clinic in Shreveport, Louisiana.*

Daniel Cutler (2012–14): "To me, the best part of being Mike VI's caretaker wasn't getting to go to football games or people being impressed by your job. The best part was getting to know Mike on a personal level where I knew him and he knew me. Everyone knows and loves him as the heart of LSU. I got to know him as a stubborn but goofy cat." *Dr. Cutler is an associate veterinarian at the Audubon Nature Institute in New Orleans, Louisiana.*

Macy Trosclair (2012–14): "There are so many moments throughout my two years as caretaker that I will cherish for the rest of my life. One stands out. During my fourth year of veterinary school, I was working very long days while on the internal medicine rotation. I would get up at 4:30 each morning to get Mike VI ready for the day before heading into the clinic, and then see him again in the evening to bring him into his night house, usually having to return to the clinic afterwards to finish up paperwork. During the last days of that rotation, one of my long-term patients was doing poorly and had to be euthanized. It was my first time losing a patient that I had worked so closely with, and at that stage in my career, I was still learning how to deal with the weight of that emotionally.

"I went to the night house to feed Mike that evening, emotionally and physically drained from the events of the day, and sat on the floor waiting for Mike to come inside. He snuck up behind me, rubbed his head against the bars, chuffed loudly in my ear, and licked my hand as if to say, 'I know this is hard, but it will get better. You got this.' He sat on the floor at my side for almost thirty minutes, completely ignoring his dinner until I was ready to get up and move on with my night. It was a simple gesture, one most pet owners might overlook if their dog or cat had done the same, but it was what I needed in that moment, and I'll never forget it.

"Being one of Mike's caretakers was a dream come true for me. It's the reason why I chose to attend LSU for veterinary school and part of the reason why I wanted to become a veterinarian in the first place. When I was a child, my family would attend most home football and baseball games, and no trip was complete without a visit to Mike's enclosure. He has always been the anchor of that campus, keeping a firm tether on the rich traditions and unwavering spirit of LSU. Even now, Mike represents the best of what it means to be an LSU alumnus. He's empathetic but strong, playful but fierce, and intensely loyal. He is the heart and soul of this university, and I am immensely grateful to have been a part of his life." *After completing a zoological medicine internship at North Carolina State University, Dr. Trosclair practiced wildlife medicine in Florida before moving on to a small-animal practice in Natick, Massachusetts.*

Lindsey (Clemones) Carr (2014–16): "Caring for Mike VI will always be the highlight of my veterinary career. He is so much more than just a tiger or just a mascot. He is the heart and soul of LSU and an important part of campus history. I was so lucky to be a part of that." *Dr. Clemones works at Dodge City Veterinary Hospital in Denham Springs, Louisiana.*

Trevor Davis (2014–16): "My favorite memory of Mike VI was when we returned Mike to his habitat after his radiation therapy planning. He was still sleepy from the anesthesia and lying on his bed. I walked up to the door and crouched down to his level and said in a soft voice, 'Hey buddy.' He gave me a sleepy chuff and I knew then that he was safe in recovery." *Dr. Davis owns Randolph County Veterinary Clinic, a mixed-animal practice in Pocahontas, Arkansas.*

Caitlin Bell (2016–18): "As a Louisiana native, I knew that caring for Mike VI was a once-in-a-lifetime opportunity. I also knew what an honor it would be to care for a college mascot. One of my favorite things

Mike VI running into his pool. (Michael Westphal)

about working with Mike VI was that Mike knew me! There is nothing like daily being up close and personal with such a large and exotic animal. Mike is not only the face of LSU; he is the spirit behind the fans and the students of LSU. Mike has a special place in LSU history. I will always treasure my time with him." *Dr. Bell is a small-animal practitioner in West Monroe, Louisiana.*

Christian Folk (2016–18): "As I lived in Baton Rouge for most of my life, visiting Mike V was a staple of my childhood. My mom used to bring my older sister and me to visit him pretty frequently, so I guess the desire to be his caretaker began there. I did not realize until right before I started veterinary school that our class would be in line for the position, but once I found out, I wanted the job. Within the first week of our very first semester, Caitlin and I talked about wanting the position and decided to apply as a team when the opportunity arose.

"My favorite thing about getting to know Mike VI was his personality. In reality, he was a pretty big goofball. He did not like being woken up early in the morning. He would play possum if he wanted you to think he was asleep and just be left alone. Taking care of him was just like getting to know any new animal. It was an amazing experience to watch him slowly over time trust you and respond to you. Seeing twenty to thirty people around him at the habitat and he's not paying anyone any individual attention, then you come by and say, 'Hey, buddy,' and he instantly pops his head up and stares right at you. It's an incredible feeling.

"Mike is the face of LSU, but with his cancer diagnosis, I think he also became a new face for hope. Watching him recover from his radiation therapy and go about his days like nothing was wrong meant a lot to me and several others. The love and support that was shown to Mike VI after his diagnosis was made public was overwhelming in the best way. He meant a great deal to many people near and far. Being able to care for such a loved and admired animal made this job extremely rewarding."

Christian recalls that Mike VI would often lay his head on the entrance door into his night house out of spite when he didn't want to come in at night, or he would partially lie within the squeeze chute so Christian or his partner couldn't trap him inside the chute. He also recalls that Mike VI would often cuddle with his big purple boomer ball when sleeping or have it in his bed with him.

Christian gained a new appreciation for Mike's ability to remember his former caretakers. Many of them came to visit him in his last few months of life after being gone for years. Christian observed that as Mike VI saw his former caretakers, he appeared to immediately recognize them and chuffed his greetings.

Christian remembers that Mike VI was a delicate eater; that is, he tended to eat over the course of the night as opposed to eating all at once. Finally, Christian recalls that of all the scents the students put out in Mike's enclosure as a form of environmental enrichment, his favorite was Vicks VapoRub. He seemed to really like the smell of the menthol and other oils! *Dr. Folk is a small-animal surgery resident at Cornell University in Ithaca, New York. Following his residency, he intends to practice small animal surgery in the Northeast.*

CHAPTER 6

SHOULD LSU HAVE A LIVE TIGER ON CAMPUS?

People often ask why it is important or valuable for LSU to have a live tiger mascot on campus. If you ask any member of the LSU community, you will get a different answer, but common themes emerge from these responses. Many people love having a tiger because of symbolism—that is, what a tiger is and represents, both within and outside of the athletics program. After Mike I assumed his responsibilities on October 24, 1936, just three days after arriving on campus, he traveled to Shreveport to help LSU defeat Arkansas 19–7. There was a rousing half-time show in Mike's honor, and everyone noticed an increased excitement for LSU football. Gordon Doré, a member of the LSU Board of Supervisors, attended that first game. He recalled, "It was during the Depression days. Everyone needed anything that they could get to sort of pump them up a little bit. . . . The fact that our university was going to have a live tiger was absolutely fantastic! It was good for the university and good for the state." Many years later, Chancellor Mark Emmert summarized Mike's contribution to LSU by saying, "Mike the Tiger is quintessentially LSU. Innately wise, fiercely competitive, ungrudgingly proud, and ever majestic, in living form Mike represents the enduring character of all people and things LSU." For many people, Mike the Tiger represents everything good that they remember about their time at LSU. That is a powerfully emotive symbol.

Mike VII looking regal. (Katie Miley)

On the morning that Roscoe was to be released into his yard, thereby becoming Mike VI, more than one thousand people waited to see the new mascot. I spoke with one family from Kentucky who had been preparing dinner the night before when they learned that Mike VI would be introduced

Fans of Mike V gather to see his return from the zoo to his new enclosure in 2005. (Ginger Guttner)

the next morning. Without hesitation they turned off their stove, got in their car, and drove to LSU. Another family from Texas said that upon receiving the notice, they finished their dinner and drove to LSU. Many people considered the tiger's release into his yard to be a historic moment they wanted to share with their children.

For others, having a tiger on campus is a reminder of our role as stewards of the natural resources of our community, state, country, and planet. Having a live tiger on campus increases our awareness about wildlife conservation, pollution, and other global challenges. In 2011 and 2012, LSU created interactive tiger education displays with support from a Department of Housing and Urban Development grant. The first display is located in the Louisiana Museum of Natural History at LSU and incorporates the mounted pelt of Mike I. The second is located in Alex Box Stadium. These displays allow thousands of guests per year to learn about Mike and about the conservation of tigers and other endangered species. In this way, Mike the Tiger shares in the university's educational mission.

For many campus visitors, Mike's habitat serves as a favorite meeting place for family and friends, and visiting Mike the Tiger provides opportunities to build and strengthen special relationships. Many adults have fond memories of visiting Mike with a parent, grandparent, or other family member. My own story is of a friend whose mother was dying. As a small token of his appreciation for her, my friend wrote down sixty memories of simple actions his mother had taken for which he was very thankful. He had these professionally inscribed by a calligrapher, matted, and framed, and he presented the gift to his mother, who was then in the hospital. Toward the top of the list he had written, "Thank you for taking me to see Mike the Tiger." Those Sunday afternoon tiger visits had in some small way strengthened the special bond between a mother and her young son.

Among the values of having a live tiger in residence is the personal connection that many seem to feel with Mike the Tiger. I had not anticipated the impact that Mike VI's illness and death would have on members of the LSU community. Many people who were themselves battling cancer, or who had a friend or loved one facing that struggle, were personally affected by the news. The day that Mike's treatment plan was announced to the public, one woman posted a message to her friend on Mike's Facebook page, "See, you and Mike can fight cancer together." After the death of Mike VI, when I visited the habitat to check on the ongoing habitat renovations, several individuals approached me and told me how much it meant to them that LSU had taken such great care of Mike VI because they too were struggling with cancer at the same time. Some had even been diagnosed the same day as Mike. Know-

lower left and right Students frequently visit the tiger education display at the Louisiana Museum of Natural History at LSU. (*left:* Valerie Stampley; *right:* Eddy Perez, LSU)

below Tiger education display in front of Mike's enclosure.

ing that somebody cared about cancer in a tiger gave them hope in their own struggles.

Not everyone thinks that a tiger's place is at a major university. Some believe that Mike should be "returned" to the wild, not realizing that none of LSU's tigers ever lived in the wild and that they would not likely survive if released to fend for themselves. Also, it is worth noting that wild tigers are exposed to viral, bacterial, fungal, and parasitic diseases that captive tigers don't have to contend with. And captive tigers don't need to hunt and kill their food, nor must they fight and defend mates and territory. It's a hard life out there for a tiger, which is why the average life span of a wild tiger is only about 8–12 years, while captive tigers typically live 14–18 years.

Below I've listed some other positive aspects of housing a live tiger on the LSU campus. You are encouraged to come to your own opinion on the issue. No one of the items below is sufficiently compelling to justify the keeping of a large exotic animal on campus, but together they demonstrate considerable educational value in a live tiger mascot.

1. **Student Education:** Caring for Mike the Tiger provides educational experiences for LSU veterinary students, adding to their medical training and helping form them into more balanced professionals. Also, tiger care, management, conservation, and regulatory matters are taught in veterinary school courses, and in meetings with veterinary student and undergraduate student organizations, so that many LSU students learn from our having a mascot.

LSU School of Veterinary Medicine faculty discuss the care of Mike V.

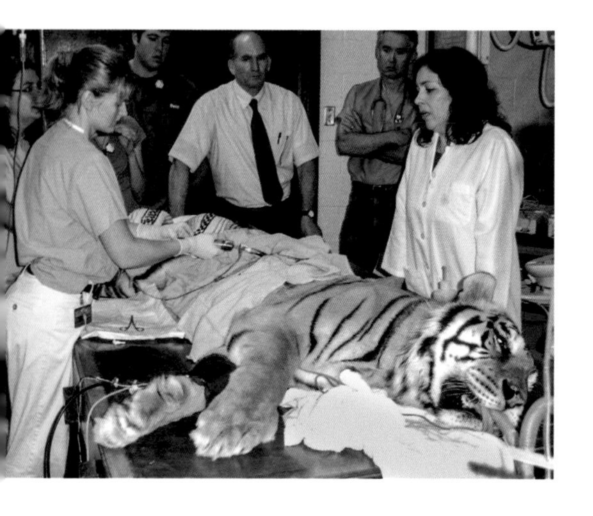

2. **Advancements in Veterinary Medicine:** Having Mike the Tiger on campus facilitates the training of faculty from the School of Veterinary Medicine. They can then share their knowledge with veterinary students and others. Also, LSU faculty occasionally publish manuscripts in the scientific literature describing aspects of captive tiger care that are based on the care of Mike the Tiger. For example, experience gained caring for Mike VI after his cancer diagnosis allowed LSU faculty to publish a scientific manuscript describing his illness and treatment.

3. **Community Education:** As mentioned earlier, two educational displays on Mike the Tiger and tiger conservation have been con-

Mike's night house allows veterinary students an opportunity to observe Mike close up. (Katherine Seghers)

structed and are on display on the LSU campus. In addition, LSU has joined with other tiger mascot universities (Clemson, Auburn, Missouri) in the Global Tiger Initiative. Led by Clemson University, the consortium represents a collaborative partnership of like-minded land-grant universities to engage academic communities worldwide and to address issues related to wild and captive tiger conservation through research, learning, and service missions.

4. Conservation Awareness: Having a tiger on campus makes better global citizens through instilling an appreciation for conservation and stewardship of natural resources, including endangered animals. I regularly speak to citizen groups on issues of tiger care, conservation, and the mascot program.

LSU's Mascot Program

Currently, fifty-six universities in the US have live animal mascots. Most of these are dogs or horses, but there is the odd bird, razorback, bison, cow or steer, bear, and of course, tiger, in addition to a few other species. Mike is the only tiger living on a university campus. Interestingly, Mike is one of the few mascots owned by his university. Nearly all the other live animal mascots are privately owned, many of them by families who have owned multiple

Mike V takes a drink from his pool.
(Harry M. Cowgill)

generations of mascots. These families have contractual agreements with their universities whereby they agree to take the mascot to designated events representing the school; in return, the university pays for some or all of the costs of maintaining the animal.

Another feature that sets LSU's mascot program apart is that a veterinarian oversees the program. Close veterinary oversight is essential when caring for a large exotic animal such as a tiger.

Within the Southeastern Conference, "virtual" relationships have formed between the mascots. For example, when Mike V had one of his root canal procedures, Uga V, the bulldog mascot of the University of Georgia, sent Mike a get-well card. Now that's unusual; a dog wishing a cat well! And when Mike V died, Tusk II, the razorback mascot of the University of Arkansas, sent a bouquet of black roses. That was somewhat ironic because if given the opportunity, Mike V might have eaten that pig!

Many aspects of LSU's mascot program have changed considerably since the days of Mike I. There are now many more laws and regulations regarding acquisition, transport, and care of a live tiger. While Mikes I–III were purchased, Mikes IV–VII were donated. For many good reasons, it is very difficult to purchase an endangered animal. Moreover, having a tiger donated to LSU supports the current emphasis on providing a home for an animal that truly needs one. LSU does not want to contribute to the irresponsible breeding of tigers by purchasing a tiger from a breeder.

Some ask why we have not obtained our tigers from municipal zoos. Because not everyone believes that a university should own a tiger, regardless of how well cared for that tiger might be, zoos must consider public perception regarding the fate of their own tigers. Finally, since the goal is to provide a home for a tiger that needs one, and zoo tigers already have homes, it is preferable to obtain a tiger from another sanctuary needing to downsize its population.

Mike the Tiger has created lifelong memories for generations of LSU students, staff, faculty, and campus guests. Mike's caretakers will continue to proactively guide the evolution of the mascot program with Mike's well-being of foremost importance. In so doing, they will provide him with loving care, allowing future generations the opportunity to create their own special stories of Mike. The front of the tiger enclosure has recognition spots available for many more tigers. It would be wonderful if all of those spots were someday filled. Imagine the memories! Geaux Tigers!